SPACE: THE UNKNOWN REGIONS

BOOKS 1, 2 & 3

SEARCH FOR LIFE IN SPACE
NASA, UFOS, ALIENS
EDWARD KANE
SPACE MYSTERIES AND WONDERS 2020's
New Discoveries & Explorations
Edward Kane

Space 2020's: What's Up There?
Edward Kane

INTRODUCTION

"Space – The Unknown Regions" chronicles the top space missions and discoveries of the 2020's. It combines three of my most recent books "Search for Life in Space", "Space Mysteries and Wonders 2020's" and "Space – What's Up There?" into one book.

The fascination with the possibility of new forms of life existing in space goes back to Ancient Greece and Plato and continues in the 2020's with NASA leading the way. The book includes NASA's unprecedented technology being deployed to search for signs of life on Mars and throughout the vast reaches of space. I examine UFO sightings caught on tape by US Navy pilots recently released by the US Department of Defense. The book showcases the big upcoming space missions including the NASA and ESA Blackbird Mission to the Sun, the Dragonfly mission to the Titan Moon, plans for sustainable lunar bases, satellite probes of exoplanets, new stars, new galaxies and new worlds. And the book details the fascinating new discoveries that are unfolding, including SuperEarths, a planet raining diamonds and a star that astronomers say could be older than the Universe itself.

If you are interested in the latest, important space missions and discoveries, "Space – The Unknown Regions" is the perfect information source for you.

TABLE OF CONTENTS

BOOK 1: "SEARCH FOR LIFE IN SPACE"

BOOK 2: SPACE MYSTERIES & WONDERS 2020'S

BOOK 3: SPACE 2020'S – WHAT'S UP THERE ?

AUTHOR'S BIOGRAPHY

Ed Kane created and serves as Executive Producer of CEO Global Foresight, a national news/public affairs program on PBS focused on breakthrough innovations. Ed is a licensed stock broker who worked for Smith Barney. He is the author of 23 books on the latest innovations. Ed is a science graduate of the University of Pennsylvania.

BOOK 1: SEARCH FOR LIFE IN SPACE

BOOK 1 SEARCH FOR LIFE IN SPACE

INTRODUCTION

Is there life and radically new forms of life in the universe surrounding us? The collective opinion of many top global scientists and space experts at NASA, the European Space Agency and the Chinese Space Agency is YES! There are new forms of life in the universe, waiting to be discovered. Space agencies around the world are putting billions of dollars in space science, massively powerful telescopes and space missions behind their search for life in space.

NASA Director Jim Bridenstine has made finding life in space a top Agency priority. He says "we are well on our way to discovering life on another world". "NASA, UFOs, ALIENS" is a top priority mission for NASA in the 2020's. Once discovered, the aliens from space will likely be brand new, radically different forms of life. Chinese space scientists theorize the aliens may include superior civilizations with technology far more advanced than ours.

As a journalist, I've written a book "Search for Life in Space" that documents the monumental scientific undertakings now underway to make our civilization the first to find life in the universe. These endeavors have taken the question of the existence of UFO's and of extraterrestrial life – including visitors from space - from the realm of science fiction to science. In spring of 2020, the Pentagon released three videos shot by US Navy pilots that document UFOs over both the Atlantic and Pacific Oceans. The Pentagon said that it released the images to demonstrate that UFOs are real and to underscore the UFOs shot by the Navy pilots are still unidentified.

The question of whether there is life beyond Earth goes back to ancient Greece. Great thinkers like Plutarch chronicled mysterious, unidentified flying objects in the sky. Christopher Columbus, during his journey to discover America, documented in his ship diary a celestial event that he saw. Great scientists and world leaders like Sir Winston Churchill and Stephen Hawkings wrote about their beliefs that we are not alone in the vast universe. The perspectives of these extraordinary thinkers are included in my book about the search for life in the universe.

The book showcases the many global space discoveries of planets and signs of life, including Super-Earths and exoplanets with the potential to sustain life. Also showcased are the many exciting missions going to the Moon, Mars and deep into space searching for signs of life that existed long ago and searching for forms of life that exist today.

The space journeys, exploration and science being used to find new life in the vast universe around us are extraordinary and exciting. My book takes you on a journey of scientists and UFOs, of visitors from space and of the unprecedented scientific pursuit of new life forms in space.

BOOK 1 – TABLE OF CONTENTS

AUTHOR'S BIOGRAPHY

Ed Kane created and serves as Executive Producer of CEO Global Foresight, a national news/public affairs program on PBS focused on breakthrough innovations. Ed is a licensed stock broker who worked for Smith Barney. He is the author of 23 books on the latest innovations across industries. Ed is a science graduate of the University of Pennsylvania.

1. Einstein, Musk and Hawking on ET's

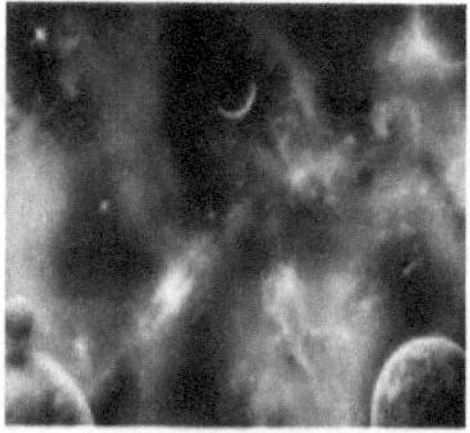

Source: NASA

Extraterrestrial Life

Since the dawn of time, humanity has wondered "Are we alone in the Universe?" Evidence is starting to mount that we are not alone, including videotapes from the US Department of Defense of UFO's or what the Pentagon is calling "Unidentified Aerial Phenomena". We begin our journey in quest of evidence of extraterrestrial life with the expert perspectives of three of the most brilliant scientists of modern times: Albert Einstein, Elon

Musk and Stephen Hawking.

Albert Einstein:
When asked if he believed in extraterrestrial life, Einstein replied:

"There is every reason to believe that Mars and other planets are inhabited," answered the professor. "Why should the earth be the only planet supporting human life? It is not singular in any other respect. But if intelligent creatures do exist, as we may assume they do elsewhere in the universe, I should not expect them to try to communicate with the earth by wireless [radio]. Light rays, the direction of which can be controlled much more easily, would more probably be the first method attempted."

Elon Musk:
"… if there are super intelligent aliens out there, they're probably already observing us. That would seem quite likely and we are just not smart enough to realize it."

Stephen Hawking:
"There is no bigger question. It is time to commit to finding the answer, to search for life beyond Earth."
"If aliens visit us, the outcome would be much as when Columbus landed in America, which didn't turn out well for the Native Americans," Hawking says. "We only have to look at ourselves to see how intelligent life might develop into something we wouldn't want to meet."
"We don't know much about aliens, but we know about humans. If you look at history, contact between humans and less intelligent organisms have often been disastrous from their point of view, and encounters between civilizations with advanced versus primitive technologies have gone badly for the less advanced. A civilization reading one of our messages could be

billions of years ahead of us. If so, they will be vastly more powerful, and may not see us as any more valuable than we see bacteria."

2. Pentagon Releases Images of 3 UFO Incidents

Source: US Department of Defense

Navy Now Has UFO Policy for Navy Pilots

We live in very strange, out of this world times. First, the global pandemic hits with unrelenting fury across the world. Now there is evidence of Unidentified Flying Objects over the Atlantic and Pacific Oceans spotted by US Navy pilots. The images come from none other than the US Department of Defense. The images have been made public. The Pentagon did so to show that the incidents are "real" and that the UFO's are still unidentified.

Three Documented UFO Incidents

The Pentagon released still photos and unclassified videos of three separate UFO incidents taken by US Navy pilots. One incident took place in 2004, involving 2 Navy jets and a US Navy ship. The other two incidents happened in 2015.

2004 "Unidentified Aerial Phenomena"

The Pentagon is calling the incidents Unidentified Aerial

Phenomena. The 2004 sighting occurred 100 miles out over the Pacific Ocean. Two Navy pilots videotaped a flying object 40 feet long and 50 feet above the ocean. As the Navy planes approached, the UFO made a rapid ascent. According to one of the pilots, "accelerating like nothing I've ever seen before." The pilots left the area and started flying to a rendezvous point 60 miles away. Their Navy ship radioed to them that the UFO was at their destination point. It had traveled 60 miles in 1 minute.

2015 Incidents

The other two incidents that occurred in 2015 involved swarms of UFO's racing across the sky and rotating in mid-air. Five pilots reported seeing the incidents over the Atlantic Ocean
from Virginia to Florida. The US Defense Department released the images to document the existence of Unidentified Aerial Phenomena. The US Navy has issued policy and instructions for their pilots to follow, should they have a close encounter with ET and UFO's.

3. MIT Discovery Expands Search for Life In Space

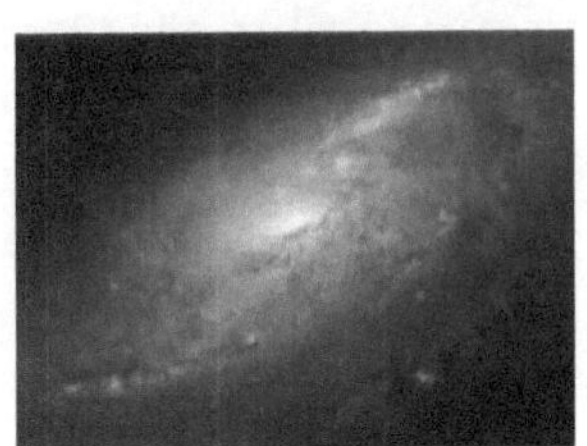

Source: NASA's Hubble Telescope

Life Found in Inhospitable Atmosphere – 100% Hydrogen

Scientists at MIT have made a major discovery. They've found that microorganisms - life - exists in an atmosphere of 100% hydrogen. In fact, organisms found on Earth can survive in environments previously considered inhospitable and composed of 100% hydrogen.

Expanding the Search for ET

The findings have a big impact on the search for life in space. A significant number of exoplanets, or orbiting stars, contain atmospheres with large amounts of hydrogen. These exoplanets have been ignored in the search for alien life until now. Award-winning astrophysicist MIT Professor Sara Seager, who led the study, is urging astronomers to expand the types of planets where they are searching for ET.

Warm Planets

Prof. Seager also discovered that worlds in space, even at a distance ten-times away from a star with a hydrogen-rich atmosphere, would be capable of containing life because of hydrogen's warming effects. Hydrogen make the planets warm. She says, it makes Super Earths - planets similar to Earth but up to 10 times bigger - prime candidates for alien life.

MIT Research Conclusions

The MIT scientists have proven that life survives and thrives in a hydrogen world. Life could exist in worlds whose atmospheres were previously thought too harsh and hostile to sustain life. And alien life may be present on many more planets than scientists thought.

4. **Amazing Look at Jupiter's Strong Storms**

Source: NASA Jupiter Thunderstorms

Place in the Universe Where the Weather is Never Pleasant

The image is one of the highest resolution, infrared views of the Planet Jupiter ever taken. And, the view is simply amazing. Jupiter is lite up in huge thunder storms. This image is the result of the most highly advanced space technology and the scientists behind it teaming up. NASA's Hubble Space Telescope, the Gemini Observatory in Hawaii and the Juno spacecraft came together to explore the fiercest storms in the solar system. The location of some of the Universe's worst weather is Jupiter, more than 500 million miles from Earth. It appears weather is never pleasant there.

Stormy Weather

A team of researchers led by U-CAL Berkeley's Michael Wong are using the images coming in to learn how Jupiter's turbulent atmosphere and weather work. Storms are constant on Jupiter and much more powerful than on Earth. The thunderheads measure forty miles from top to base. That's five-times the average on Earth. And the lightning has three times more energy than the largest Super-bolts on Earth. The Juno spacecraft is picking up radio signals from the powerful lightning to map the bolts.

Jupiter Has Deep Water Clouds

Hubble, high in space and Gemini, on the ground, are getting high resolution images of Jupiter from afar to provide a wider perspective. They are correlating lightning storms with deep water clouds to start estimating how much water is in Jupiter's atmosphere. In the search for life in space, the presence of water is the key. Jupiter's harsh weather may not preclude it from harboring some forms of life.

Search For Life

MIT astrophysicists have discovered that life can exist in extremely inhospitable conditions. They found microorganisms living and thriving in environments that are 100% hydrogen, where scientists never even considered the possibility of life. Many rocky exoplanets in the universe have that type of atmosphere. The MIT team is urging scientists to greatly expand the search for life in space. Jupiter could be another interesting candidate. For life in space, the sky is the limit.

5. UFO Sightings on the Rise

Source: Stock Image of UFO

The National UFO Reporting Center Keeps the Count

The number of UFO sightings in North America jumped to 6,000 in 2019. That according to the nearly

fifty year old National UFO Reporting Center, whose
mission is the collection and dissemination of objective
UFO Data. It's been doing so since its founding
in Washington State in 1974.

Low flying, Extremely Fast and Silent Aerial Vehicles   The
rising number of incidents in the sky range from flashing white
lights, fireballs, disc-shaped vehicles, cone shaped vehicles,
fleets of speeding UFO's in perfect V-shaped formations and
much more. In most cases, the UFO's are flying low, incredibly
agile, totally silent and extremely fast. A number of witnesses
have reported UFO's doing very rapid ascents and then totally
disappearing. In mid-2020, the US Department of Defense released three videos show by US Navy Pilots of "Unidentified
Aerial Phenomena" or UFO's that they
spotted over the Atlantic and Pacific Oceans.

Ufology
The science of exploring and collecting data on UFO is
called Ufology. It's practiced by The National UFO
Reporting Center and by UFO expert Scott Waring, among
others. The Center reports that UFO sightings are on the rise.
In 2019, California was number 1 with 485 cases. Florida was
number 2 with 385 cases and Washington State was number 3
with 222 cases. Many of these cases were officially filed with
The National UFO Reporting Center and can be accessed
there.

6. UFO Sighting Daytona Beach, Florida

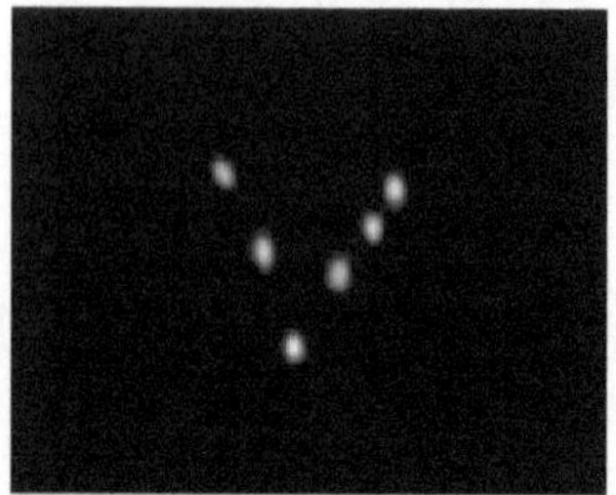

Source: University of Syracuse Sighting of V-Formation UFO

DATELINE: UFO Sightings, Daytona Beach

At 10PM (EDT) on April 6, 2020 in Daytona Beach, FL, a former "highly trained" member of the US Coast Guard reported to The National UFO Reporting Center, along with US government agencies, that he saw nine small white lights arranged in a V-formation flying directly above his house.

They travelled from the southern sky toward the Northeast. The witness said the lights appeared to be independent of each other and not attached on a single object. The event lasted for about five seconds.

Credible Observation

The witness of this UFO event is a highly trained former member of the US Coast Guard. He's a certified Raytheon 64X band radar technician. There was a full Moon in Daytona Beach that night. The witness says the V-formation was ¾ of the size of the Moon. In his report filed with The National UFO Reporting Center, the witness says the V-formation was low-flying, totally silent and extremely fast. He also notified the Federal Aviation Administration and the Daytona International Airport control tower. The nine vehicles in V-formation we're not picked up on radar.

Corroborated by The National Center

This sighting is of great interest. The Center spoke to the witness several times and found this to a highly credible sighting.

7. NASA'S SEARCH FOR EXTRATERRESTRIAL LIFE

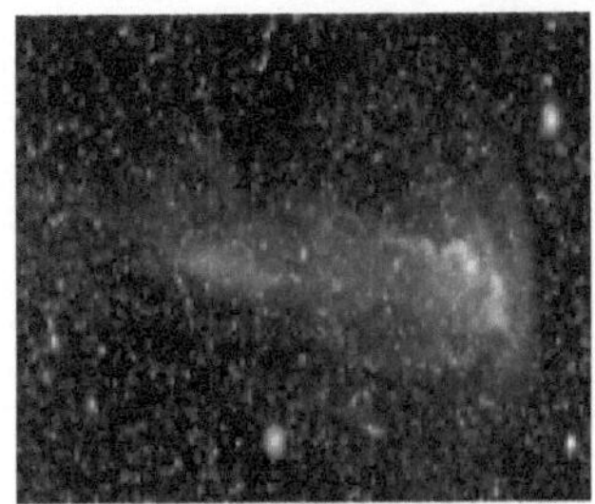

Source:: NASA

NASA Leadership and Famous Astronauts

NASA had made the search for extraterrestrial life in space a primary goal and mission. Here are the expert perspectives on the prospects from several famous astronauts and NASA's current Chief:

NASA Chief Jim Bridenstine

On NASA's objective of discovering aliens Bridenstine said:
"The goal is to discover life on another world; that's what we're trying to achieve. And because of so many great people in this room, friends, we are well on our way to doing that."

Astronaut Buzz Aldrin

After three spacewalks as pilot of the 1966 Gemini 12 mission, as the lunar module pilot on the 1969 Apollo 11 mission,
and as one of the first two humans to land on the moon, he said the following: "There was something out there that was close enough to be observed... sort of L-shaped."

"There may be aliens in our Milky Way galaxy, and there are

billions of other galaxies. The probability is almost certain that there is life somewhere in space. It was not that remarkable, that special, that unusual, that life here on earth evolved gradually, slowly, to where we are today."

Astronaut Scott Carpenter
As a NASA Project Mercury astronaut he said:
"At no time, when the astronauts were in space were they alone: there was a constant surveillance by UFOs."

Astronaut Gordon Cooper
As one of the original Project Mercury astronauts, Gordon Cooper testified before the United Nations saying this:
 "I believe that these extra-terrestrial vehicles and their crews are visiting this planet from other planets…"
"I did have occasion in 1951 to have two days of observation of many flights of them, of different sizes, flying in fighter formation, generally from east to west from Europe."

8. UFO Sighting Over Bakersfield, California

Dateline: UFO Sighting - California Hotspot
On April 15, 2019 in Bakersfield, California, a man
and his girlfriend reported that they saw a UFO

formation in the sky. He reported and filed a report with
The National UFO Reporting Center: "We just witnessed three
unidentified objects, one in the middle being the
largest, escorted by two other smaller unidentified
objects. One to the left and one to the right." He
said they were "dark colored objects heading
southeast and travelling at an unknown speed and
disappeared into the clouds beyond our sight"

Another UFO Unidentified and Unexplained
This is yet another report of an unidentified aerial phenomena
or UFO reported to authorities. This one from
California. It is one of 6,000 in the past year in
North America.

9. International Space Station and UFO's

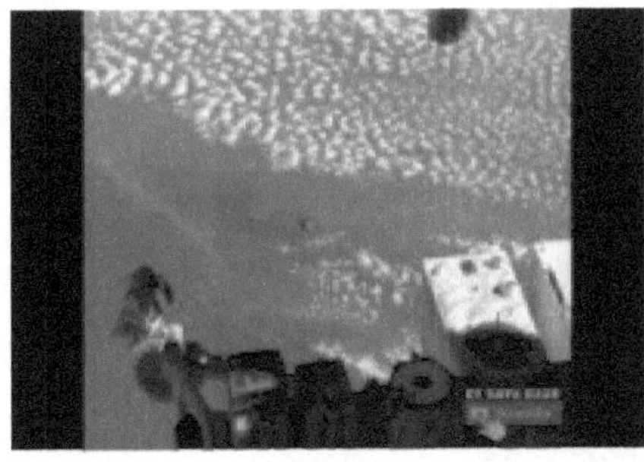

Source: ET Database

UFO Spotted Keeping Pace with ISS
This is one of several, recent incidents involving the International Space Station (ISS) and UFO's. It occurred in February
2020. ISS cameras picked up a strange, cone-shaped object in
orbit in space. The object followed and kept pace

with the ISS for 21 minutes. At one point, NASA cameras aboard the ISS zoomed in on the flying object, strongly suggesting that NASA was aware of the mysterious flying object chasing the ISS.

UFO the Size of a Bus
According to UFO expert Scott Waring, the object was the size of a bus. To keep pace with the ISS, it would have to have been travelling at 7.8 kilo-meters per second. Waring says NASA is baffled by the object. Waring adds that NASA doesn't know what the object is or why it was there.

Spy Space Plane or UFO?
This space vehicle could have been another sighting of a UFO. But some space experts offer another, unconfirmed explanation. It could have been the US Air Force's X-37B. The small, autonomous plane conducts intelligence surveillance from orbital positions in space. The X-37B can stay aloft for months and its missions are tightly classified. The truth of the matter is we really don't know what the flying object keeping pace behind the ISS was. It's yet another mysterious flying object in the skies.

10. UFO Sighting Found on NASA ISS Feed

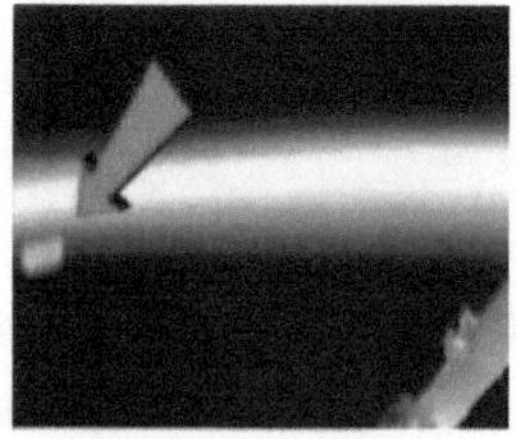

Source: ET Database

Unidentifiable and No Explanation

UFO hunters believe they've made the find of a lifetime. It's the image of a UFO "uncloaking" in front of the International Space Station. The flying object appeared from out of nowhere and took on a cylindrical shape for a brief period of time.

UFO Expert

UFO expert Scott Waring was the first to spot it during a live NASA International Space Station (ISS) feed in April 2020. Waring says that just as the object came into view, NASA shut-off the feed transmission and went to a red screen. With the live cameras shut down, the opportunity to watch the mysterious flying object abruptly ended.

UFO Shields Malfunction in Sunlight

According to Waring of ET Database, the object is very solid. The sunlit side of the Earth coming into view "caused the UFO's shields to malfunction for a short time. That's why we saw it."

Looking for UFO's?

Expert Scott Waring says if you want to see a UFO, go watch a few sunsets. He adds: "I'll promise you will see one. But bring a camera or your friends won't believe you." As for NASA, the space agency has not explained why the feed abruptly ended or what the unidentified flying object could have been. Yet another

mysterious and unidentified UFO hurtling through space.

11. China Searching for ET

Source: China's FAST Telescope

China's FAST Radio Telescope

China's FAST Telescope has started its search for alien life in space. FAST is a 500 meter-diameter Aperture Spherical Radio Telescope and it is expected to be a major force in the search for and research on extraterrestrials in space. It's the world's largest single-dish radio observatory. It's located in southwest China's Guizhou Province. It started operation in January 2020 and has discovered 44 pulsars so far.

Targets: Space Alien Civilizations

The FAST team of scientists say the telescope has the potential to detect Earth-like civilizations deep in space on thousands of exoplanets. They say they will use FAST to explore whether our neighboring Andromeda galaxy, M-31 Andromeda, has "more advanced civilizations and technology than we do on Earth".

SETI Missions: Search for Extraterrestrial Intelligence

The first SETI "Search for Extraterrestrial Intelligence" is underway. Li Di, the Chief Scientist for FAST, says first results have laid a solid foundation to continue the mission to find
new worlds with new forms of life and civilizations in the Universe around us.

12. **Moonshots for Bezos and Musk**

Source: Blue Moon

Humans Back on the Moon by 2024
NASA awarded contracts to three companies to build space-craft capable of landing astronauts on the Moon. Blue Origin, the space company founded by Amazon's Jeff Bezos, SpaceX founded by Tesla's Elon Musk and Dynetics, a subsidiary of the IT company Leidos, are the winners.

Is There Life in Space?
The contract is a big win for Bezos and Blue Origin. He has been urging NASA to explore the Moon's south pole where scientists have discovered water in the form of ice. Water is the key ingredient that NASA targets in its search for life in space. Bezos has been pushing his Blue Moon lunar lander.
He's pulled in Lockheed Martin, Northrup Grumman and Draper Labs as part of his Blue Moon development team. According to Bezos, "We're not going back to the Moon to visit. We're going back to stay." Blue Origin received the biggest contract: $579 million.

NASA's Search Goals

NASA wants to set up a permanent base on the south pole of the Moon. NASA wants to explore how the ice might sustain life and if there are any forms or remnants of life in the ice. Also, NASA wants to determine if the ice could be converted into rocket fuel for spacecraft in space.

Lunar Competition

SpaceX, Blue Origin and Dynetics will be competing against each other for the spacecraft development. The plan is to fly the astronauts in the Orion spacecraft, built by Lockheed Martin. Orion would go into lunar orbit where it would meet up and dock with the new lander spacecraft, which would fly the astronauts to the Moon.

2024 Moon Landing

NASA is pressing ahead with its goal to put humans on the Moon in 2024. It would be the first time since US Astronauts landed on the Moon in 1972. The lunar mission is called Artemis. Before takeoff, the US Congress must approve a budget for it, estimated at $35 billion to $50 billion.

Elon Musk's Vision of Space

With his company SpaceX, Elon Musk wants to move humanity deep into space. He plans to start with a base on the Moon and eventually have a self-sustaining city on Mars. Musk has enjoyed a series of NASA wins. On May 27, SpaceX will fly US astronauts to the International Space Station. He also won a $7 billion contract to ferry supplies to the ISS.

13.   IS IT A UFO?

Source: US National Forest Service

Or, Is It Not? Then What?

NASA believes there is extraterrestrial life out in the Universe and has missions, technologies and plans designed to find new life in space. Here's a picture in NASA's universal puzzle of discovery and it raises the question of what is really out there?
An intriguing UFO appeared in the sky recently over Mt. Shasta, California in February 2020. The sighting is incredibly unusual and it was initially identified as a UFO. In this case, no one is sure. But the most likely explanation is it is a lenticular cloud hovering over Mt. Shasta in Siskiejon, California. But, the oddity is no one really knows.

So, Is It an Odd Cloud?

The image shot by the US National Forest Service is incredible. It was taken at 7AM (PST) on February 12. 2020. The US Forest Service categorizes it as an unusual
weather formation - a lenticular cloud. It is beautiful, mysterious and intriguing. What do you think? Lenticular clouds are stationary, shaped like a lens and form on downward slopes of mountains. They shape with moisture particularly in winter and spring. This one looks a bit different. In the image, it appears to be poised to move forward. But there are no definitive answers on whether this is a cloud or a UFO cruising through our skies. It is another incredible image that unveils what is in the Universe and showing its presence here in our skies.

14. **Europe's ET Hunting Spacecraft**

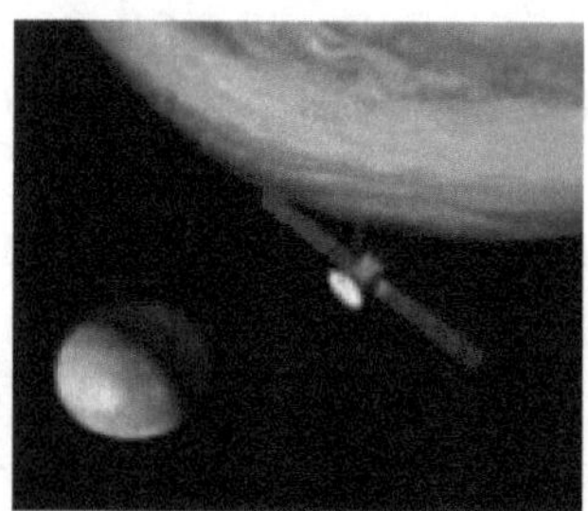

Source: ESA - JUICE

JUICE'S Search for Signs of Alien Life
The European Space Agency (ESA) is putting the final touches on a new spacecraft specifically designed to search for alien life. The first place that it will search for extraterrestrial life is Jupiter and its Moons. The spacecraft is called JUICE or Jupiter Icy Moons Explorer. JUICE's purpose is to find alien life or determine if the planet and its Moons are capable of sustaining life. This is yet another example of how seriously global space agencies, are taking the search for life beyond Earth.
NASA with its upcoming Mars probe is leading the exploration along with ESA.

JUICE's Timeline
There is a timeline for JUICE. First in Germany, JUICE is being outfitted with highly advanced communications systems, navigation sensors and onboard computers. Next stop is The Netherlands, where it will go through rigorous testing. JUICE's

launch is set for June 2022.

Underwater Alien Life

ESA is targeting for exploration by JUICE three of Jupiter's 79 Moons. The reason is the 3 Moons have the perfect environments to support water and oceans. ESA scientists
are especially interested to determine if those oceans hold alien life forms in underwater environments.

15. CHINA SOARING INTO SPACE

Source: China's Space Agency

Exploring Space for Signs of Life

China has launched its largest rocket, carrying a next generation spacecraft onboard. The Long March-5B rocket carrier took-off from the Wenchang Space Center in the southern island province of Hainan. The rocket is massive: 54 meters long and has a takeoff weight of 849 tons.

Life in Outer Space

China has big space ambitions and plans. It wants to equal the US and Russia as a space power by 2030. The experimental spacecraft is part of China's efforts to shuttle astronauts to its future space station for manned space exploration. A big component of the exploration is searching for life in
outer space. Chinese space officials have explicitly stated that they are deploying massive technologies to find what they call "new civilizations with superior technology" in space. China expects to have a multi-module space station for astronauts/

space explorers by 2022.

16. **LASER GUIDE STAR**

Source: European Southern Observatory

Laser Beams Over the Milky Way

The technologies behind new discoveries in space are amazing. An example: the European Southern Observatory's Very Large Telescope (VLT). In the night skies over Chile, the VLT beams its laser guide star. The beam of laser light from the laser guide star stretches over the Milky Way Galaxy. Astronomers are using giant laser beams to correct their highly powerful telescopes from distortions caused by turbulence in the Earth's atmosphere. The turbulence causes stars to appear to twinkle when we on Earth star-gaze at night.

Space Discoveries from Chile

The laser guidance system for powerful, advanced space telescopes discovering new exoplanets, galaxies, stars and life outside of Earth and on other Planets is based at the Laser Guide Star Facility at the ESO's Paranal Observatory in Chile. Some of the most powerful telescope systems in the world for peering into space are located at the Observatory.

Search for Habitable Planets

Paranal is also home to a huge, 4 telescope system called SPECU-

LOOS or Search for Habitable Planets Eclipsing Ultra-Cool Stars. The system watches for dips in the brightness around "ultra-cool" stars, similar to brightness dips when the Moon eclipses the Sun. The image below is the view SPECULOOS has provided of the Carina Nebula in deep space. The Carina Nebula is a nursery for baby stars.

Source: ESO image of Carina Nubula

17.NASA & Actor Tom Cruise Making Movies in Space

Source: NASA - ISS

International Space Station & a Big Star

This is innovative film making at its best. NASA is working with actor Tom Cruise to shoot a movie in space. Specifically, NASA has disclosed that it will fly Tom Cruise to the International Space Station (ISS). Cruise will stay aboard and shoot a movie there. And NASA Director Jim Bridenstine says this isn't science fiction. It's going to happen! Apparently, Tom Cruise is about to become an astronaut.

ISS

The ISS is a multibillion science lab that orbits 250 miles above the Earth. Rotating crews of astronauts have been onboard since 2000. The US and Russia are the primary operators. According to Director Bridenstine, NASA's top priority is to find new forms of life in space. He is putting billions of dollars of new technology, missions and telescopes behind the effort in the 2020's. It will be interesting to see if the Cruise movie plays into the UFO and ET themes.

First Movie Shot in Outer Space

Cruise's movie will be action adventure aboard the ISS. It will have the distinction of being the first Hollywood movie shot in outer space. It isn't clear when Cruise and his crew will be launched. But in a tweet, Bridenstine said NASA needs popular media to inspire a new generation of scientists and engineers to help make NASA's "ambitious plans a reality." The Cruise movie has the potential to be a blockbuster that could be of great benefit to NASA.

18. NASA's Massive Solar Mission of Discovery

NASA's Image of Solar Particle Storm Spewing Plasma into Space

Six Tiny Satellites Form One Massive Telescope

NASA's is using unprecedented technology for discoveries in its next journey to the Sun. NASA has plans to deploy the largest radio telescope to ever fly into space. The system contains six tiny satellites that work in sync, fly in tanden and come together like a tech rock band to form the world's biggest telescope in space. The goal is to investigate powerful Solar storms. The mission is called SunRISE, or Sun Radio Interferometer Space Experiment. It is designed to help scientists significantly upgrade information on the Sun's dangerous activity around the Earth called Space Weather. NASA hopes to launch the satellite system in 2023.

Space Weather to Help Track the Journey

Understanding and predicting space weather events are very important because of future plans for expeditions to discover new life in space and for ordinary citizens to start travelling into space to the Moon, Mars and beyond. The new satellite radio telescope system will start in a low Earth orbit. NASA's goals are to understand the Sun and how it erupts with space weather events. The hope is experts will learn how to mitigate the potentially dangerous effects of space weather on space travel, spacecrafts, astronauts, exploring for life in the universe and for all travelers in space.

19. Black Hole Catching a Star

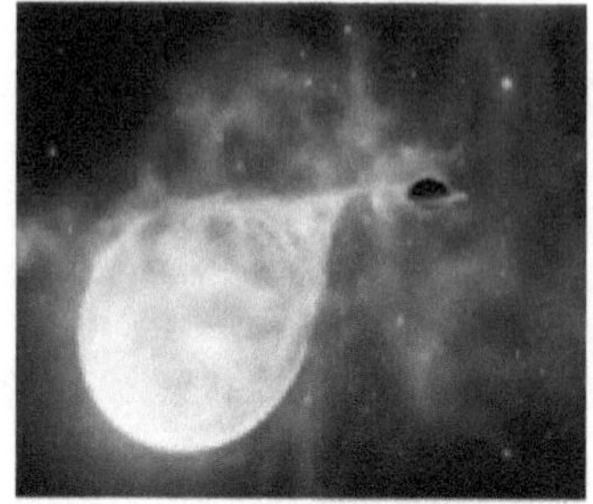

Source: ESA/.Hubble Artist Rendering

Elusive Missing Link of Intermediate Massive Black Holes

The Hubble Space Telescope has detected what looks like a very elusive intermediate-mass black hole in space (IMBH). The Hubble picture captures the black hole in the process of pulling in a passing star. Incredibly, the black hole is 50,000 time the mass of the Sun. If this discovery proves to be true, it will be a breakthrough. It will provide astrophysicists a big jump forward with the missing link they've been looking for to understand the evolution of black holes. They don't understand how intermediate sized black holes assemble and if they are the building blocks for massive black holes. This is another example of the dynamic universe that surrounds us. It's another journey into the search for life in the universe beyond the Earth and the beginnings of the Universe itself.

Tidal Disruption of a Star

Hubble, in its remarkable space journey, caught this image in April 2020. Astrophysicists at the University of New Hampshire in Durham have published their analysis. They describe the Hubble image as the "tidal disruption of a star" by an intermediate-mass black hole. The action took place in a massive star cluster at the outskirts of a huge galaxy. This galaxy is way beyond our Milky Way. According to NASA, this happened deep within space in a distant, dense star cluster.

Why Is This Important?

Why is this discovery so important? Because it has been so elusive and so critical to understanding the evolution of black holes. Intermediate-mass black holes are smaller, less active and much more difficult to detect then massive ones in space. And astrophysicists believe they are a key link in the chain of back-hole creations in the Universe, that go back to the beginning of life itself.

20. Most Powerful Supernova Ever: Star Explosion

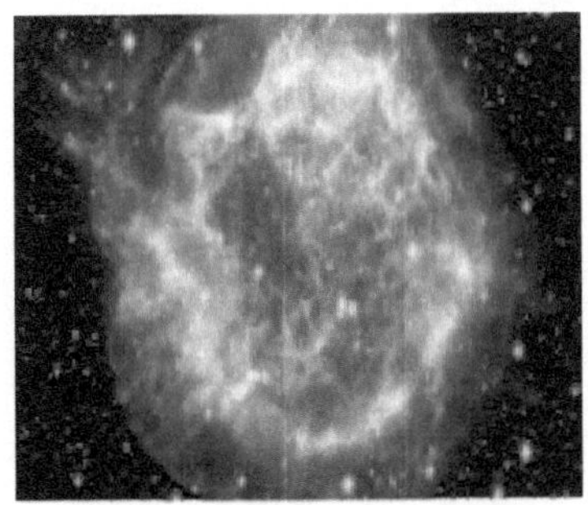

Source: NASA

Spectacular Space Event
An international team of scientists has discovered the most powerful and brightest supernova ever recorded. The space event is off the charts. A supernova is the explosive death of a star. In this case, scientists theorize it may have been two stars that merged and then exploded. The supernova event is unique,
extreme and one for the record books. It takes us back to the creation of life, the Universe and stars. The mysteries of life itself and the existence of life in space were likely contained in this supernova.

Billions of Light Years Away
The event occurred 3.6 billion light years from Earth. It's named
SN2016aps. According to lead astronomer Matt Nicholl of the University of Birmingham, the team measured the supernova in

two ways: total energy of the explosion and the amount of light or radiation emitted. With normal supernovas, the radiation amounts to 1% of the total energy. In this one, the radiation is an incredible 5 times the explosion energy.

Bigger than the Sun

This supernova produced the most light astronomers have ever seen. They believe it contained 50 to 100 times the mass of the Sun. Their report on the discovery has been published in the journal Nature Discovery. The astronomers say they believe even more exciting discoveries on the evolution of life and life in the universe around us are on the horizon.

21. New Discoveries for Hubble

Source: NASA, ESA Hubble

Space Missions Ahead

The Hubble Telescope has been exploring and discovering new

worlds in space for thirty years. It was deployed in space on April 25, 1990. More than anything else, it has showed mankind the Universe around us. Its current mission is to probe deeper and deeper into space as part of NASA's search for signs of extra-terrestrial life.

Early Years

The first few years of its space tour of duty were riddled with technical problems, particularly with its mirrors. But new technology and new cameras were launched and deployed by astronauts on Hubble.

Hubble Field

By 1994, it began its riskiest and richest mission – the Hubble Field. For ten days scientists used Hubble to examine what seemed like a dark spot in the Universe. No signs of any brightness, probably no stars. What Hubble found were thousands of distant stars and galaxies. Those findings, back in 1994, have forever changed our view of the Universe, as an immense world of discovery.

What's Next for Hubble

Over the years, Hubble's journeys have moved deeper into space for wide field, celestial surveys like Hubble's Ultra Deep and eXtreme Deep Fields. They've revealed very distant, deep in space massive galaxy clusters. With new instruments, Hubble's future missions include WFIRST and LU-VOIR which will take the spacecraft deeper and deeper into space. It is on rocky exoplanets outside our solar system that many astrophysicists and scientists expect to find new forms of life. Hubble's legacy and its future is to bring the Universe into closer focus for all of us to discover.

22. WRONG WAY METERORITE

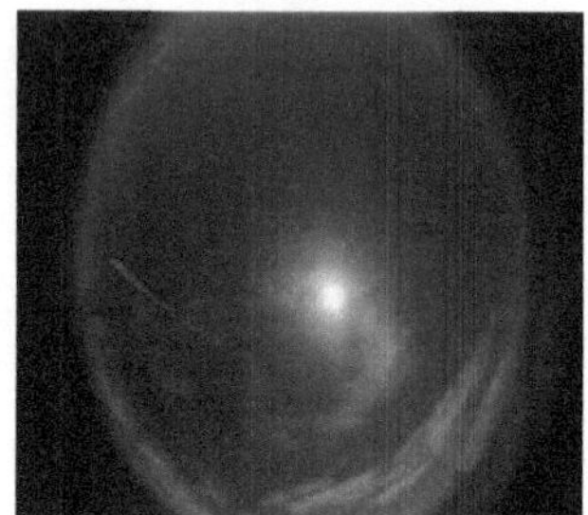

Source: Australia's Desert Fireball Network

Streak of Light the Width of Texas
The celestial event, called a Grazing Fireball, happened over
Western Australia. The meteorite that blazed over the night
skies took a highly unusual path. For a minute and a half it
burned, carving an arc of blazing light as wide as the state of
Texas, then, it faded away and took back off into space. It's
being called the meteorite that ignored the one way signs
of our Solar system. Meteorites that crash and splinter on Earth
are a favorite target for scientists to look for the basic building
blocks of life in the Universe.

On to Jupiter
What's incredible is this meteorite didn't disintegrate in our atmosphere or crash to Earth. After the light display faded, it
took off back into space. According to Australian astronomers
and scientists, who released their findings in spring 2020 on the
event, that happened in July 2017, the wrong way meteorite is
headed to Jupiter They say, the likely arrival date is 2025.

Blazing Speed
This is an extremely unusual celestial event. And, it's an extremely rare meteorite that grazes into the Earth's
atmosphere at a very low angle. It skipped across the skies like
a skipping stone on a pond. The Australian team estimates that

the meteorite weighs 130 pounds, is a foot across and moves at a blazing speed of 10 miles per second. They believe it originated in the asteroid belt between Jupiter and Mars.

Ignoring the One Way Signs of Space

Grazing fireball events are extremely rare. In 1783, historical records tell us that that the Great Meteor streaked across the skies of England and Europe and streaked back to space. There was a similar meteorite, the "Great Comet", that blazed across the skies of the northeast US in 1860. And one was spotted in 1972.

Interstellar Space

When the Australian meteorite reaches Jupiter in 2025, it will have to grapple with Jupiter's gravity. Then it is likely to be tossed into interstellar space. Astronomers are not speculating on the next leg of its quixotic journey. This is the first grazing fireball that's been studied by scientists. And, it's also the one that spent the longest amount of time in the Earth's atmosphere, as far as scientists know.

23. NASA CLOCKS MASSIVE WINDS IN DEEP SPACE

Source: NASA

Winds on a "Failed Star" of 1,450 MPH

NASA has achieved yet another first in space. NASA scientists have clocked winds of 1,450 mph on a huge "brown dwarf" or failed star, that is 33.2 light years away and outside of our solar system. The object is the size of Jupiter, the largest planet in the solar system, and 40 times Jupiter's mass. The winds are extremely fierce. This discovery is a first for astronomers. Previously they were only able to measure winds speeds on planets and objects within our solar system. Even though the atmospheric conditions are extreme, NASA is greatly expanding its search for new forms of life deep into space outside of our solar system on space objects like this.

Whole New Worlds

The scientists say this discovery opens the way to understand atmospheres unlike anything in our solar system. The technique they used to make this discovery is also a first. The discovery was made by combining the detection of radio and infrared emissions from the brown dwarf. They measured the spinning rotation rate of the dwarf and the rotation rate of the atmosphere around the dwarf. By comparing those rates, the NASA scientists made history and establish the wind speed around a huge object in distant space.

24. Are We Alone in the Universe?

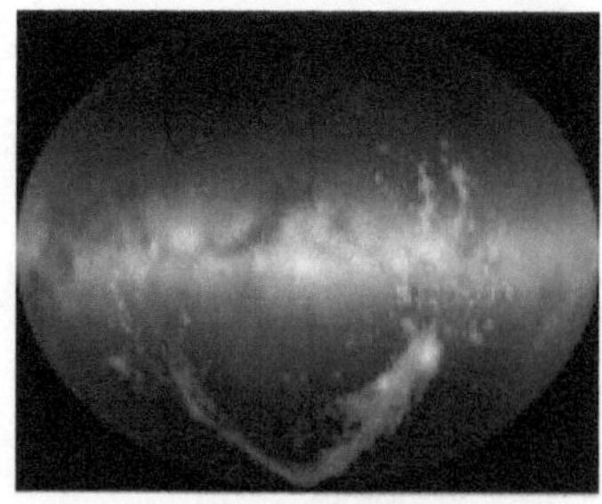

Source: NASA

Humans have been searching the skies and wondering if we are alone in the Universe since the dawn of time. Here are three sightings going as far back as the 1st century AD from Plutarch,

Thomas Jefferson and Christopher Columbus

Plutarch

Ancient philosopher and biographer Plutarch (A.D.49 -119) chronicled an incident during the battle between Lucullus and Mithradatis:

Plutarch wrote, "when he faced (Lucullus) the enemies, he was amazed by the number of people and wanted to avoid and delay the battle. As Marios who was dispatched from Iberia by Sertorius to help Mithridatis as a sergeant together with a military army challenged him (Lucullus), he got set for battle. While marching towards the conflict, without any observable change happening, the air opened and appeared a rapidly descending object resembling a flame, which appeared like a vase in shape and like a glowing annealed metal in color. Both armies, frightened by the sighting, withdrew. They said that this happened in Phrygia, near to what is known as Otries."

Thomas Jefferson

In 1800, soon before he was 3rd President of the US, Thomas Jefferson wrote about a UFO sighting in Louisiana by William Dunbar to The American Philosophical Society: It was apparently "the size of a house," Jefferson wrote, and the "color of the sun near the horizon. Immediately after, it disappeared in the north east, a violent rushing noise was heard, as if the phenomenon was bearing down the forest before it, and in a few seconds, a tremendous crash was heard similar to that of the largest piece of ordnance causing a very sensible earthquake."

Christopher Columbus

Italian explorer and colonizer Christopher Columbus and members of his crew on October 11, 1492 saw a strange light which ascended and descended on his trip to America. It was logged in the ship's diary. Columbus wrote: It was like "a small wax candle that rose and lifted up, which to a few (crew members) seemed to be an indication of land." His son Ferdinand also described it as akin to a candle that went up and down.

25. Red Fan Skies Over Japan

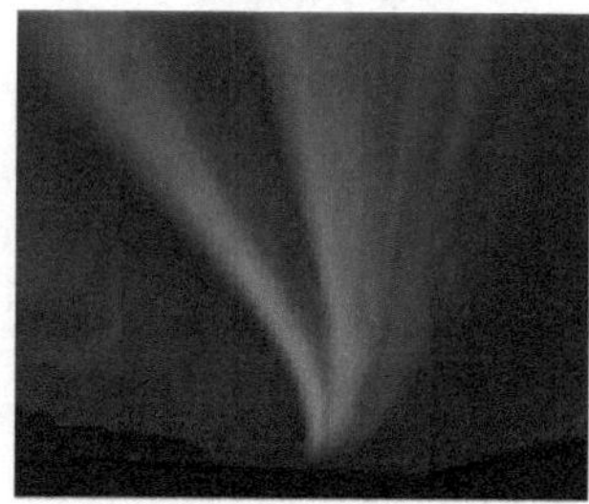

Source: Space

1400 Year Old Mystery

More than 1400 years ago, a dazzling red fan spread out across the skies over Japan, according to historical records. The phenomenon has been documented by modern day astronomers and has baffled them ever since. For the inhabitants of Japan in the year 620, the red skies were frightening. They called it a "strange red sign" in the sky. Was it another form of life in the universe? Perhaps a threatening form of life? Astronomers and space weather experts, working at the National Institute of Polar Research in Japan, have just determined that the red skies were in fact what they call "an ethereal sky phenomenon."

Pheasant Tail

The scarlet fan-like light over Japan happened on December 30, 620. Those in Japan who saw it said it looked like a pheasant tail and many considered it a bad omen, something threatening coming from the skies. For years, modern astronomers have considered it to have been a comet or an aurora. But, neither seemed to fit until now.

Space Weather
Japanese astronomers have now solved the mystery. The explanation is space weather and the 7th century heavenly display is a spectacular aurora. This is a space event in which elements of the Earth's atmosphere are being activated by charged particles spit out by the Sun. Astronomers have also discovered more recent displays of auroras over Japan that are pheasant tail shaped and scarlet red like the aurora in 620. The 1400 year old space mystery has been solved in 2020.

26. Famous Brits and Aliens

Source: Winston Churchill.org

Have you ever wondered what British Prime Minister,
Sir Winston Churchill, Britain's first astronaut and
the UK's leading UFO expert for the Ministry of Defense
think about the possibility that extraterrestrial forms of life exist in the vast realm of space? They are all believers. And Winston Churchill's perspectives go back to a 1939 essay that was just found.

Winston Churchill
Sir Winston Churchill wrote an essay on alien life in 1939: "I,

for one, am not so immensely impressed by the success we are making of our civilization here that I am prepared to think we are the only spot in this immense universe which contains living, thinking creatures," he wrote in the newly uncovered essay, "or that we are the highest type of mental and physical development which has ever appeared in the vast compass of space and time."

Helen Sharman

Helen Sharman, the first British astronaut in space, said this to the Observer: "Aliens exist, there's no two ways about it. There are so many billions of stars out there in the universe that there must be all sorts of different forms of life," she went on. "Will they be like you and me, made up of carbon and nitrogen? Maybe not."

Nick Pope

Nick Pope, a top UK UFO expert and investigator for the Ministry of Defence said: "It has been a widely held belief in Ministry of Defence circles that 'aliens' have been able to detect us for decades via TV and radio broadcasts. What once seemed like science fiction is steadily being realized by central governing bodies as distinctly real. If aliens have studied our psychology, they may choose to appear in our skies on a significant date – the closing ceremony of the Olympic Games is one date being widely circulated by conspiracy groups."

27. Famous Americans and Extraterrestrials

New Forms of Life in Space
The existence of life outside of the Earth is not a matter
of science fiction. Top US government officials including two
Presidents - Ronald Reagan and Richard Nixon, as well as FBI Director J. Edgar Hoover are among those
who have had an extraterrestrial experience.

Richard Nixon
President Nixon's remarks during a news conference:

"I'm not at liberty to discuss the government's knowledge of
extraterrestrial UFO's at this time. I am still personally being
briefed on the subject." Jackie Gleason, a close friend of the
president, was reportedly taken to Homestead Air Force Base
where Nixon showed him the bodies of extraterrestrials. Gleason's former wife Beverly McKittrick said that he described the
aliens as small beings that are "only about two feet tall with
bald heads and disproportionately large ears."

Ronald Reagan
President Reagan's remarks on a flight he took over California:

"I looked out the window and saw this white light. It was zig-
zagging around. I went up to the pilot and said, 'Have you ever
seen anything like that?' The pilot said no. President Reagan

asked them to follow it. "We followed it for several minutes. It was a bright white light. We followed it to Bakersfield, and all of a sudden to our utter amazement it went straight up into the heavens. When I got off the plane I told Nancy all about it."

J. Edgar Hoover
The first director of the FBI response to people spotting UFO's over Los Angelis in 1942:

"We must insist upon full access to disks recovered. For instance, in the LA case the Army grabbed it and would not let us have it for cursory examination."

28. Interloper From Another Solar System

Source: NASA 21/Borisov

Not Your Average Comet

NASA and global astronomers agree that 21/Borisov is not your average comet. It arrived from another solar system and is packed with carbon monoxide (CO) and some water, suggesting it may have formed in a cold region within its own planetary system, far away from its Sun-star. And, far away from Earth and our solar system. 21/Borisov is raising many questions and few answers. Water is the key ingredient for the existence of life.

What Is ET?

Astronomers are perplexed about what 21/Borisov truly is. Could it be an alien probe? No one really knows. Since it arrived in our skies in August 2019, astronomers have been diligently observing it. It is the second interstellar object ever discovered to have ventured into our solar system.

Oumuamua

The first interstellar object found soaring through our solar system was il/Oumuamua. It was discovered in October 2017 as it was on its way back to outer space. It too was very unlike any usual comet. Oumuamua was cigar-shaped and had no comet tail.

Best Theories

Some astronomers believe that 21/Borisov is an ancient space relic of planetary formations billions of years ago and the beginning of life itself. Exploring it would provide a window into the distant past when planets formed. And perhaps because even now that 21/Borisov contains water and CO, the possibility that it holds life or clues to the existence of other forms of life, in the Universe so vastly surrounding us, is an intriguing mystery to be explored.

29. Evidence of Ancient Life on Mars

Source: Curiosity Rover on Mars

Primitive Martian Works Of Art

UFO expert Scott Waring says he has seen proof of alien life on ancient Mars after examining photos of the Red Planet taken by NASA's Mars Curiosity Rover. The objects he has drilled down on in the photos, he says, look like ancient Martian carved rock images. Waring says the images demonstrate the existence of complex alien lifeforms that once lived on Mars and created forms of rock art. Waring runs a UFO focused blog called ET Data Base.

Images of Interest

Curiosity shot the photos in an area of the Gale Crater, a hot spot in the search for life by NASA, known as Shaler. The images were taken on SOL 315 or June 25, 2013 on Earth. This area of Mars is barren terrain with a very rocky surface. Waring spotted a strange object on the ground that looks like a carving. The carving appears to be that of an elongated face that was carved on a smooth object. It is totally different from the jagged rocks surrounding it.

Perspectives from Waring

"The smooth stone didn't fit into its rough surroundings. But the details on the stone reveal a face carved into it. It looks like some kind of a dog-like face but the cranium extends far back, making the face 3 times larger." Besides that, Waring seems to have spotted fossilized bones and teeth around the object. This is another look through technology at the many possibilities of life in the universe far back in time and far beyond the Earth.

30. Worlds Away New Asteroid

Source: NASA

Travelling at 32,000 MPH

The European Space Agency (ESA) has discovered a huge asteroid moving at warp speed. 32,000 miles per hour to be exact. The asteroid is a big one with a diameter of 262 feet. That's about the size of a commercial aircraft wing tip to wing tip. ESA has named the space rock 2020 Ft3 and they've added it to their Risk List of asteroids that could hit Earth. Remnants from asteroids are a primary resource for scientists to examine for any signs of life sustaining components in deep space.

2089 - 2110 Timeframe

2020 Ft3 follows an Earth-crossing orbit, making it a member of the Apollo family of asteroids. The space rock is worlds' away from Earth. ESA believes the chances of it striking the Earth are very small. And, they and NASA have calculated, if by some slight chance it were to hit, it would not happen before 2089. They add, the remote possibility of a strike would be between 2089 and 2110.

Asteroid Defense System

NASA and the ESA are currently working on an asteroid defense strategy to use spacecraft to collide with any threatening asteroids and bump them off course, if headed toward Earth.

31. Satellite Spots First Alien World

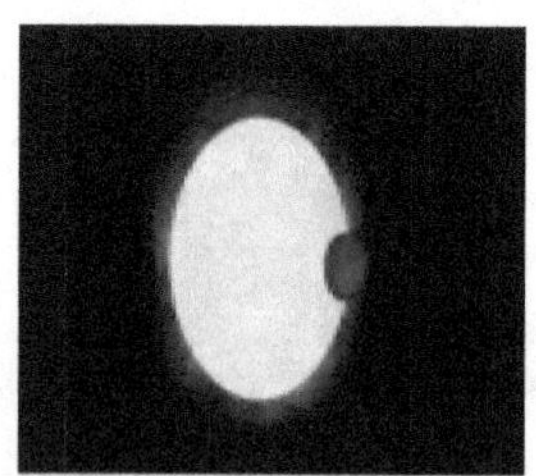

Source: ESA's CHEOPS image

ESA's New Highly Advanced Satellite

Europe's new exoplanet-spotting space telescope CHEOPS has been tested in space for the past three months. It has successfully spotted several exoplanets. These are planets that orbit stars outside of our solar system. They're known as alien planets. These exoplanets are considered by NASA and other experts as likely hosts for new forms of life. By May 2020, it will start scientific observations of great importance to the process of discovering life beyond the Earth.

CHEOPS

CHEOPS is on a different mission than NASA's Kepler space telescope that searches for exoplanets. The CHEOPS space telescope is targeting known exoplanets to determine their sizes and density. Thanks to its highly advanced instrumentation, the accuracy of measurement that CHEOPS delivers will be unprecedented.

Search for Alien Life on Alien Planets

The discoveries and findings from CHEOPS are expected to be critical to scientists in their efforts to determine what the exoplanets are composed of and whether they have the capability of sustaining alien life. On that basis, scientists will prioritize the most promising planets to explore for life.

32. NASA's TINY ROBOTIC ROVERS FOR MOON

Source: NASA

Big Science: NASA Wants Tiny Science Payloads

NASA has plans to send tiny robotic rovers to the Moon. In order to do this, NASA needs to develop miniature scientific payloads for the tiny rovers. NASA formally asked the public for their ideas and will deploy the best among them.

Game Changing Science

This is not fun and games. NASA expects the winning ideas

that come through this public process to be game changing
They believe the smaller payloads will
result in a series of miniaturized instruments and sensors that
can quickly be used for lunar exploration, including for the
search for forms of life particularly in the South Pole of the
Moon where ice is known to exist. Water is the key ingredient
to sustain life. The tiny rovers will also be used for harvesting
the resources found on the Moon.

Crowdsourcing

NASA's Jet Propulsion Lab has launched a crowdsourcing project
on HeroX. The title: "Honey, I shrunk the NASA payload". Specifically, NASA has targeted small instrument design concepts
to sustain human life on the Moon and also to use the resources
found there.

Ice Hunting Robot Also On the Way

The instrument concepts must be about the size of a Milky Way
candy bar or a bar of soap: 3.9" by 3.9" by 1.9" and weigh no more
than 0.8 ounces. This tiny rover program is additional to NASA's
water ice-hunting robot VIPER that is set to launch in 2022.

33. Planets with Double Sunsets Exist

Source: NASA Artist image of 3 planets & 2 stars in the Kepler-47 system. That's 3,340 light years away from Earth

300 Exoplanets With More than One Sun-Star

New research shows that double sunsets on planets may be just as common as the one sunset a day we enjoy on the planet Earth. Astronomers say that has important implications for the search for other forms of life outside of our Solar System. New research has discovered more than 300 exoplanetary systems that have more than one Sun-Star and double or more sunsets per day. This is Star Wars come alive, with Luke Skywalker's Home planet Tatooine showcasing its double sunsets in iconic images in the George Lucas movie series. Only now there's proof Tatooine type planets exist in the Universe.

Where There Are Twin Suns, Could There be Life?

Astronomers prioritize Sun-like stars as the most probable places in the universe where alien life could exist. For massive stars that are most like the Sun, new research tells us that 50% of them exist with companion stars. Astronomers are now focusing on these binary stars to see if they are hosting Earth-like planets. That possibility has almost been ignored until now. The change of focus happened when a team of astronomers using data and observations from Europe's Gaia spacecraft discovered more than 300 distant planets with binary Sun-stars. The game-changer is the exponential innovation in new technology to explore distant space. The search for life in these exoplanets is now starting to unfold.

34. $10 Billion Webb Space Telescope: Deepest Explorer

Source: NASA's Webb image

Hubble's Incredible Replacement Partner in Space

The $10 billion James Webb Space Telescope "Webb" will launch in March 2021. Its mission of space discovery is awesome and its technical capabilities are intergalactic. Webb will give all of us on Earth a brand new look at the massive, dynamic mysteries of the cosmos around us and accelerate the search for new forms of life in the Universe around us.

Webb's Mission of Discovery

Webb is the highly advanced successor to the amazing Hubble telescope. Webb has advanced technology to detect infrared light that enables it to look further back in time than any other telescope. It will explore the mysteries surrounding the creation of the Universe, photograph and image distant galaxies, stars and exoplanets and further understanding of the dynamics of the solar system, the creation of life and the existence of new forms of life in new worlds.

Webb is Unique

Webb is the most advanced, complex and ambitious space telescope ever developed. It's a joint venture between NASA, the

European Space Agency (ESA), the Canadian Space Agency (CSA) and the Space Telescope Science Institute. The telescope is massive and contains an infrared camera and spectrograph to collect light from the early beginnings of the Universe and of life, itself. It is able to see much deeper back into time than any other space telescope. It has the capability of imaging the dawn of the cosmos, detect first ever images of stars, galaxies and exoplanets. Its instruments can block the light of a star in order to detect distant exoplanets and their atmospheres. This will be a time tunnel journey into the beginnings of the Universe that is soon to unfold.

Building On Hubble's History
Hubble has made enormous space discoveries in the past 30 years and into 2020. By comparison, it is a $2.5 billion, 11 ton telescope launched by NASA in 1990 with a primary viewing mirror diameter of 8 feet by 2.4 meters. Webb is a 6 ton telescope with a primary viewing mirror diameter of 21 feet by 6.5 meters with greatly advanced infrared resolution. The new views from Webb will redefine the universe as we know it.

35. NASA Discovery Accelerates Question of Life Beyond Earth

Source: NASA Artist Rendering

Earth-like Planet: Kepler-1649c
Have astronomers found another Earth? A team of

international astronomers have discovered a planet the same size
as the Earth. Like the Earth, the planet orbits its Sun Star.
This exoplanet is called Kepler-1649c. The planet seems to be habitable. Could there be life there?

Habitable
Scientists say the planet seems to have habitable regions. It is orbiting in its star's habitable zone. That is a region around the star where a rocky planet could support water.

Remarkable Discovery
The scientists discovered the Earth-like planet while using re-analyzed data from NASA's Kepler space telescope. Kepler was retired in 2018. Previous scans of the data with a computer algorithm misidentified the space object. Astronomers taking a hard look recognized it to be a planet.

Space Specifics
This planet is 300 light years away from the Earth. Of all the exoplanets discovered by Kepler, this one is most similar to Earth in size and in temperature. The amount of light it receives from its star is 75% of what the Earth receives from the Sun. However one big difference is that the planet's star is a "Red Dwarf", which are known for stellar flare-ups. That could make life, as we know it at least, on the planet a challenge.

Amazing Discoveries
According to NASA Associate Director Thomas Zurbuchen,
this discovery gives us greater hope than ever "that a second Earth lies among the stars." He adds, even more amazing discoveries are on the horizon.

36. NASA Diligently Searching For ET on Mars

Source: NASA Caltech Space Image

What's Underground on Mars?

NASA scientists have a fascinating and very innovative theory. They're exploring the possibility that some forms of life could be supported underground and inside caves on Mars. They believe there's a chance that water may exist underground to support life on the Red Planet.

2020 Launch

In July 2020 NASA will launch its Mars 2020 rover with a mission to search for life on the Red Planet. It will touch down on the Martian surface on the targeted date of February 18, 2021 at a location called Jezero Crater. NASA calls the area one of the most scientifically interesting landscapes on Mars.

Search for Water and Life

Jezero Crater was once home to an ancient river delta that may contain signs of microbial life. The possibility exists that there is water underground on Mars. Exploring for water and life is rover's big mission. The rover will join lander InSight that has detected seismic activity on Mars. In fact, hundreds of Marsquakes in the past year were detected by InSight. The cost of NASA's July 2020 space mission to Mars is $2 billion. NASA's rover will be joined by the European Space Agency's ExoMars rover with a Mars landing scheduled for March 2021. Literally, billions of dollars are being spent by NASA and the European Space Agency to search for any signs of life on Mars, which they are betting is a real possibility.

37. Flying Drone, Flying Car or UFO?

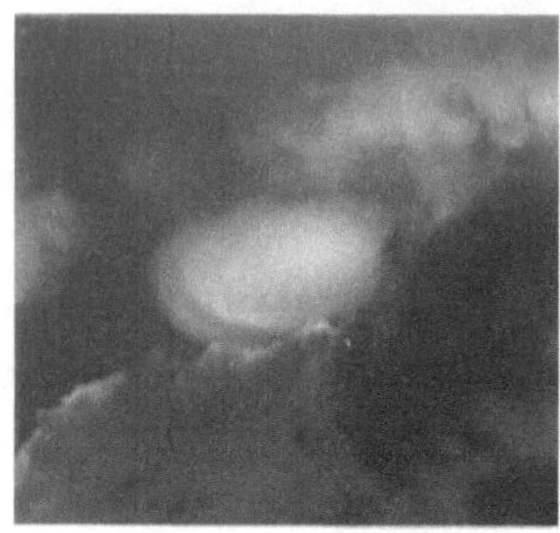

Source: Stock Image of UFO

"Alien Vessel" Caught on Tape Above a Car

This is a strange case of a fast flying object - white with a cloud-like appearance - that a driver accidentally taped on his smart phone through the car's open sunroof. He turned over the tape and reported the incident to California authorities. The incident happened in Santa Clara, California on a highway close to San Francisco Bay. The question is what was it: a flying drone, flying car or a UFO?

Caught on Tape

The video of the object can be viewed on the enclosed URL. The object moved with great speed and is only visible for a second or two. The object is flat and elongated. UFO expert Scott Waring of ET Data Base examined the tape and concluded that it is very similar to other alleged UFO's sighted over the years around the world. He adds "there's a very real probability that it's a real UFO'. He speculated that the "alien vessel" might have come from an underwater base in San Francisco Bay. This is yet another interesting sighting of an unexplained UFO's possibly from outer space.

Video
https://www.bing.com/videos/search?
view=detail&mid=500B33B728EC6DDB4BCB500B33B728EC6DDB4BCB&shtp=GetUrl&shid=72ec52c7-

144d-4b0b-
b9df-55b043c21248&shtk=R3V5IENhdGNoZXMgVUZPIE92ZXIgU2FudGEgQ2xhcmEsIENhbGlmb3JuaWa

38. NASA: Possibility of Life on Titan, Saturn's Largest Moon

Source: NASA Jet Propulsion Lab

Moonscapes, Science and Life

NASA scientists have mapped Saturn's exotic, largest Moon Titan. Apparently, Titan has quite a story to tell about the possibilities of life. NASA has unveiled the first global geological map of Titan. It shows a strange and exotic world that NASA scientists say is a strong candidate for the search for life beyond Earth.

Incredibly Interesting Terrain

The map shows dunes of frozen organic material, vast stretches of plains, and lakes and seas filled with liquid methane...all of which may harbor forms of life beyond Earth. The map is based on radar, infrared and data from NASA's Cassini spacecraft that studied Saturn and its moons from 2004 to 2017. The data and images now emerging are fascinating.

What's Next?

Clearly, Titan contains organic material that is the ingredient critical to form life. What's next is NASA is going to launch its Dragonfly Mission sometime in the 2020's to reach Titan by 2034 with a multi-rotor drone to explore this exciting new Moon in space.

39. New Food for Space & Earth to Search for Space Life

Source: Prague University Scientists

Aeroponically Grown Vertical Vegetables and Herbs

As humanity searches in spacecraft for new life forms out in the deepest reaches of space, what kind of food can they sustain themselves on? Czech scientists from Prague University have created a new lab to experiment on growing food for very constrained and extreme conditions with a lack of water like on Mars. But the extra utilization is also back on Earth to boost agricultural output in places hard hit by urbanization and Climate Change.

Marsonaut Experiment

The Czech scientists call it the Marsonaut Experiment and it's being led by their top life scientists. They're using aeroponics to grow plants in the air, without any soil and very little water. To minimize space, they grow the plants in vertical units stacked one on top of the other.

Harvest Time

The team has succeeded in growing salad leaves, mint, basil, mustard plants, herbs and radishes. They enjoyed the fruits of their harvest and said it tasted great. Their next big crop is strawberries. The scientists are focused on light and temperature changes to maximize yields. What is so interesting about this innovation is that it's not only targeted at a menu on future human journeys to Mars for explorations to sustain life and fine life. It also has applications here on Earth because the growing process uses 95% less water and saves space. The technology could make a big difference in arid areas and urban spaces. It could also sustain space explorations into deep space by humans.

40. Out of This World Exoplanet

Source: NASA Artist Rendering

Exoplanet LHS 3844b

A NASA space telescope has revealed for the first time a rocky Earth-sized exoplanet. It's outside of our solar system and tightly orbits around the most common type of star in our home galaxy, the Milky Way. The exoplanet appears to have no atmosphere to sustain life. But, with new emerging scientific discoveries on the abilities of extremely harsh places in the universes that may sustain some new forms of life, this exoplanet may also have a lot about life to reveal.

Atmospheric Void

Researchers published their discovery in the journal Nature. They say the surface of the exoplanet, known as LHS 3844b, is most likely barren like the Moon or covered in dark volcanic rock. The exoplanet is 49 light years away from the Earth. It's one of 4000 exoplanets, discovered in the past 20 years, that orbit stars in the Milky Way. It's 1.3 times larger than the planet Earth.

41. Martian Rock Tells Big Story

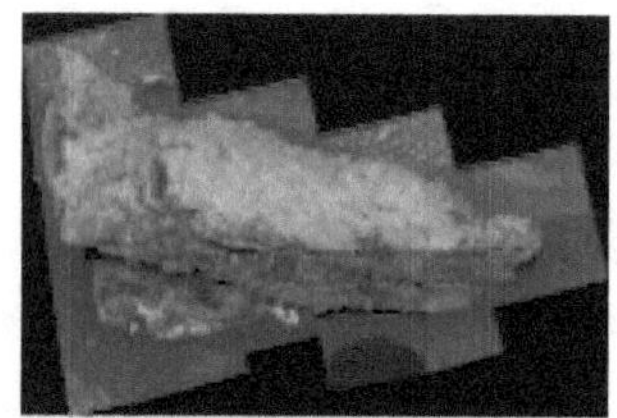

Source: NASA

Shows Presence of Water, Raises Questions of Life on Mars
NASA's Curiosity rover has been exploring the planet Mars for the past seven years. It recently found a rock that contains some of the secrets to what Mars was like when it was a lot more wet. The discovery is so important that the rock has been nick-named "Strathdon".

Strathdon
The rock was found in the Gale Crater or the "clay bearing unit". The rock is composed of layers of sediment that are unique and wavy. NASA scientists say this shows evidence of flowing water and blowing wind. They add the history of water on Mars is much more complicated than they had previously thought. They now do not believe that Mars went from wet to dry over-night.

Signs of Life on Mars
The question still remains: with the presence of water on a warmer Mars with a thicker atmosphere, could that have sus-

tained more than microbial life? NASA is searching for that answer.

42. Super-Earth Found in Space by Global Ground Telescopes

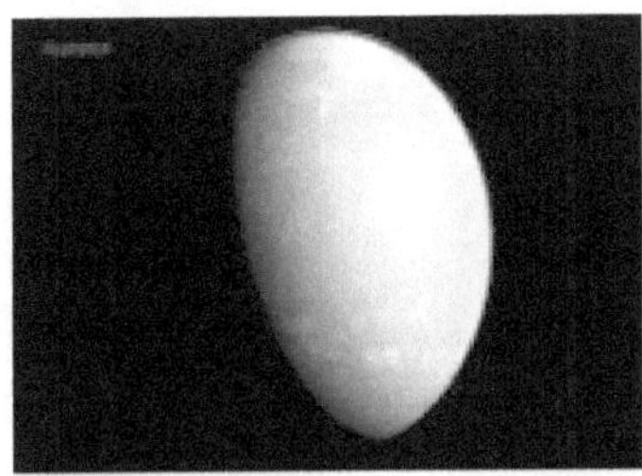

Source: NASA's Goddard Space Flight Center

May Have "Earth Like" Conditions

Astronomers have discovered what they're calling a Super Earth 31 light years away from our solar system. They say it is "potentially habitable". The planet is named GJ 357d. It is six times larger than the Earth and it orbits a sun that is much smaller than ours. An international team of scientists discovered it using powerful ground telescopes and following up on data first relayed by NASA's planet hunting satellite TESS. The astronomers believe that it could provide "Earth like conditions" for life.

Looking for Signs of Life

Highly advanced telescopes will, in the next several years, be going on line that will be able to pick-up any signs of life on Super-Earth. Two telescopes going live in 2021 and 2025 should show whether the planet is rocky and if it has any oceans. If the planet's atmosphere is thick, it could support water on the surface and sustain life. If there is no atmosphere, the average temperature would be 64 degrees below zero, making it more glacial than habitable.

43. NASA's 21 Whole New Worlds

Source: NASA

Planet Hunter Extraordinary

NASA's planet hunter TESS spacecraft - Transiting Exoplanet Survey Satellite - has marked its first year in space and the results are extraordinary. The spacecraft has discovered 21 new worlds outside of our solar system. It's also spotted 850 possible exoplanets and 6 supernovas. For NASA the results are way beyond all expectations.

Star Search

TESS is focusing on stars that are less than 300 light years away. It looks for dips in brightness which indicates that an object just passed across the star. The data will help NASA determine which exoplanets it wants to explore in-depth and where life might exist. TESS is a technology planet hunter to track down the most likely places for new forms of alien life in the Universe that surrounds us.

Historic Mission

This is the most comprehensive search for planets ever accomplished by mankind. TESS finished exploring the southern part of the sky and is examining the north for new exoplanets and new places to explore for life.

44. Sustainable Human Bases on the Moon

Source: Stock Image of Future Lunar Bases

Fascinating Research by the European Space Agency

Administrators of the European Space Agency (ESA) believe that the next step in space exploration is building lunar bases for astronauts. The purpose is deeper human exploration into the Universe to discover new worlds and new forms of life that may exist. To do that a sustainable source of energy is required to sustain human life and power for rovers and landers. ESA scientists think that they have one, using the surface of the Moon. ESA scientists have created bricks composed of lunar regolith, which is the soil, dust and rocks on the surface of the Moon collected on past lunar missions. In testing, the bricks are able to store solar energy and generate it for use as power, heat and electricity. It's possible this innovation could provide future lunar bases a sustainable energy source.

Scaling-Up the Technology

The ESA team is now scaling up the process and efficiency of their heat energy bricks from lunar regolith. They say that travelers to the Moon, with this new energy source, wouldn't have to take much with them from Earth. They also think it has the potential to enable very ambitious missions into space. As the world marked the 50th anniversary of the Apollo 11 historic mission to the Moon, the work being done by the ESA showcases humanity's continuing quest and ingenuity to journey into space, form habitats there and search for new forms of life.

45. NASA's Dragonfly Mission

Source: NASA's Dragonfly Artist Rendering

NASA Searching for Life & the Origins of Life

NASA plans to launch its Dragonfly Mission in 2025. It's an audacious plan to land the Dragonfly rotorcraft lander on Titan, which is Saturn's largest moon. It's the only moon in the solar system that has an atmosphere. According to NASA, Titan is most comparable to early Earth and, most importantly, it has the ingredients to support life. This is NASA's next mission to a new destination. It will also mark the first time that a multi-rotor vehicle is deployed on another planet.

Source: NASA - Titan

Mission That's Out of This World

Dragonfly has the capacity to fly 100 miles through Titan's thick and cold atmosphere. It will make trips on a daily basis to a number of test sites. The terrain is diverse and includes oceans, rivers, dunes, lakes and craters. Saturn and its moons are ten times farther away from the Sun than the Earth is. There is water on Titan but it is so cold, given the distance from the sun, that the water is frozen all of the time. NASA believes Dragonfly's exploration of Titan on the ground and in the air may reveal the processes that led to the origins of life on Earth and possibly even find some forms of life on Titan.

Touchdown on Titan

The Dragonfly drone will land on Titan in 2034. The journey consists of 888 million-miles. Dragonfly will search for any signs of life past or present. NASA says its instruments can evaluate organic chemistry and "the chemical signature of past and present life". There is an abundance of nitrogen and methane on Titan and NASA scientists believe there may be complex organic molecules, that could be the basis of life. It will be a fascinating journey into outer space to search for the origins of life here on Earth and in potential new forms of alien life.

46. From Meteorites, NASA Finds New Clues to Life

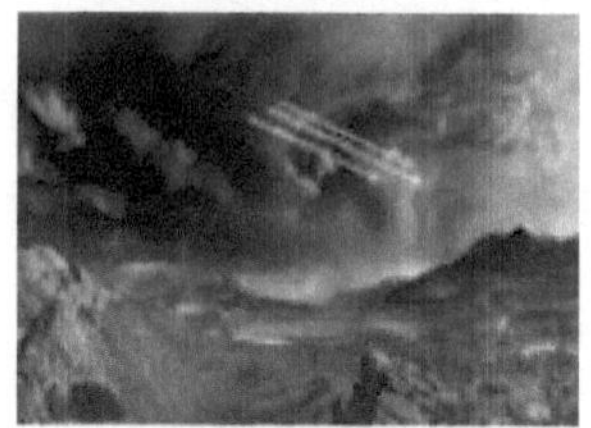

Source: NASA

Essential Molecules for Life

NASA and a team of university scientists have discovered compounds containing cyanide, carbon monoxide and iron in carbon rich meteorites from space. They think these compounds may have helped power life on earth. They've published their findings in the journal Nature Communications.

Early Earth

The team of scientists think that cyanide was probably an essential compound for building molecules necessary for life. Besides being present in meteorites, they say the compounds were also present in early Earth before life began when the Earth was constantly bombarded by meteorites.

BENNU

Data collected by NASA's OSIRIS-REx spacecraft of the asteroid BENNU will be delivered to earth in 2023. NASA scientists intend to search for the same compounds which according to NASA scientist Jason Dworkin of the NASA Goddard Space Flight Center "may have helped start life on Earth or on other bodies in the solar system". It's a fascinating innovative research and discovery journey in space to determine the origins of life here on Earth and perhaps elsewhere.

47. Tumbling Pieces of Wrecked Planets Or Alien Life?

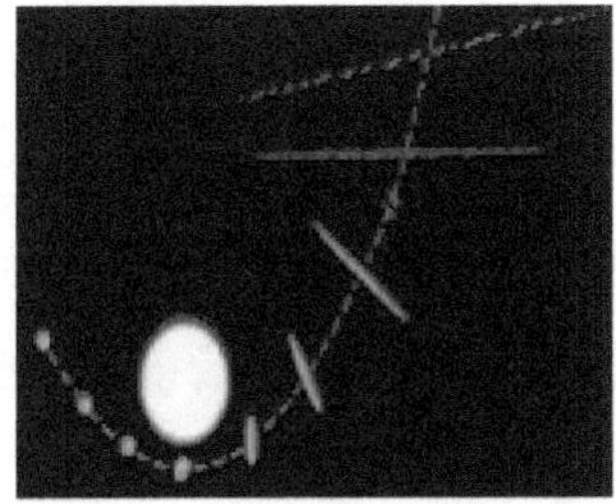
Source: NASA/ESA

Hurtling Through Space

Astronomers call the tumbling space object Oumuamua. In Hawaiian, that means a message from afar. That's what it certainly is. Oumuamua was first discovered in 2017. Some scientists thought it could be a vestige of some sort of alien life, because everything about this space object is so unusual. The question is: What is Oumuamua?

New Discoveries and Findings

There is no definitive answer as to what Oumuamua is and what it may represent in terms of new arrivals from space. But, two astrophysicists, Yun Zhang at the Observatoire de la Cote d'Azur in France & Doug Lee at U-CAL in the US, have different findings. They believe that the reddish-colored, cigar shaped object tumbling through our solar system is the fragment of a wrecked planet that got too close to the star it was orbiting around. The gravitational force of the star destroyed the planet and the fragment is among the pieces that are left.

First Arrival

Oumuamua is the first object from another star system found passing through our solar system. Fortunately, it's passing out of our solar system and will reach Uranus in August 2020. The fragment is 1/4 miles long. And, it's elongated like a cigar in shape. It has very odd motion and a dry appearance. The scientists say it isn't like an ordinary comet or asteroid. It is another unexplained aerial phenomena in the universe to be explored.

Computer Simulations

Nothing is certain yet but computer simulations tell the team of scientists that this may be the remnants of a wrecked planet, that blew up from the stronger gravitational forces of another object in space such as the "tidal forces" of a star. This is one of the extraordinary new discoveries made in the dynamic and mysterious universe all around us.

48. NASA Discovery of DNA in Space

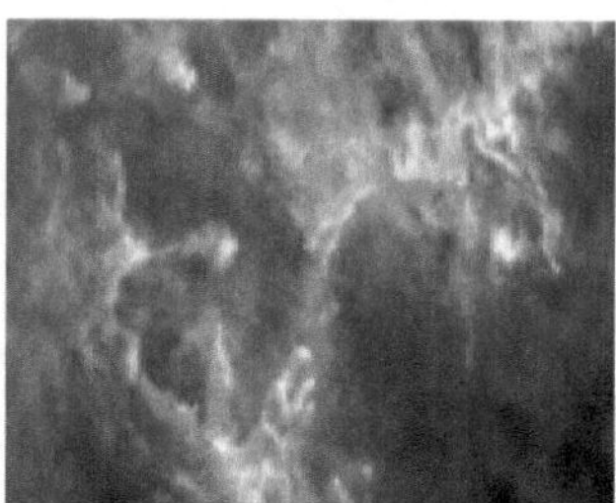

Source: European Space Agency Shot of Eagle Nebula - Frigid & Radiation Rich

Stuff of Life

Parts of DNA, the stuff of life, can form in space. NASA scientists have made deoxyribose, the sugar that is the backbone of DNA, under space-like conditions. In a lab, they blasted ice with radiation and discovered 2-deoxyribose. Their innovation and findings are published in the journal Nature Communications.

NASA Research

NASA astrochemist Michael Nuevo says their research shows that the process of DNA formation can happen anywhere in our galaxy. It suggests that the stuff of life could have been delivered to earth from elsewhere.

Process
The scientists cooled frozen water and frozen methane to -260 degrees. Inside a vacuum, they blasted it with ultraviolet light mimicking conditions in interstellar clouds. Warming the irradiated ice simulated what occurs when a young star is born. The scientists identified 2-deoxyribose in the ice.

Asteroid Missions
NASA and Japan have two asteroid missions going on. They will bring back samples. And the scientists hope to search for deoxyribose in them.

49. Bigger Than Big Bang

Source: NASA

Huge Explosion in Universe From Light Years Ago Detected
Astronomers using data from the most advanced NASA, European Space Agency, India and Australia telescopes have discovered the biggest explosion that the universe has ever experienced since the Big Bang. It happened in a galaxy 390 million light years away from the Earth. And incredibly it took place in slow motion over hundreds of millions of years. The science to discover this phenomena is remarkable and the details emerging on events going back into space from such a distance and time are amazing, innovative scientific discoveries.

Supermassive Black Hole in the Ophiuchus Galaxy Cluster
A global team of astronomers say the explosion they detected released five times more energy than ever detected before. The explosion came from a supermassive black hold in a galaxy called the Ophiuchus Galaxy Cluster nearly 400 million light years from the Earth. The power of the explosion was enough to punch a hold into the gas surrounding the black hole that is big enough to contain 15 contiguous Milky Way galaxies. This is amazing scientific discovery through breakthrough telescope innovation. What it means to the formation of the Universe, Earth, life and new forms of life are the next chapters of the history of the Universe.

50. Encounters Between Navy Aircraft and UFOs

Source: stock image of UFO

Incident Filings by Pilots
Several Hazard Reports released by the US Navy document a number of sightings and encounters by Navy pilots with unidentified flying objects or UFOs. The Hazard Reports were released under a Freedom of Information request. They document six separate incidents. This is in addition to the three videos released by the US Department of Defense in spring 2020. The videos and images were shot by Navy pilots encountering incredibly fast and agile "unidentified aerial phenomena" over both the Atlantic and Pacific Oceans. The UFOs were spotted by

at least seven Navy pilots and a US Navy ship.

Official UFO Sightings by US Navy Pilots
The Hazard Reports were officially filed by six Navy pilots during the past decade. The pilots, in their official reports, said the UFOs encountered looked like missiles, unmanned drones, while some of the other aerial objects were smaller. In at least one of the incidents, Navy pilots could not get radar to lock on and track the object.

UFO, Drone or Balloon?
In another incident that occurred in 2013, the pilots thought the UFO might be a drone, but no operators could be located. The pilots said the wingspan of the object was 5 feet. It was white in color with "no distinguished features." One of the pilots followed the object for an hour off the coast of Virginia. Two other reports released by the Navy involve balloon-like objects in April 2014. One of the balloons almost caused a mid-air collision with an A-18 F Super Hornet. These are the latest US government reported incidents of UFO sightings with the added credibility of US Navy pilots filing official reports.

BOOK 2

SPACE MYSTERIES AND WONDERS 2020's

INTRODUCTION

In my new book "Space Mysteries and Wonders 2020's", I showcase the latest big discoveries in space like a planet raining diamonds, a Super-Earth that could contain life, the mysterious Planet X and a star that's older than the Universe itself. We chronicle new, mysterious objects just spotted in space that scientists have never seen before, spectacular galaxies that shouldn't exist and the birth of baby stars and planets.

We also examine innovation in space technologies like the world's first Mars helicopter, the super-secret USAF Space Plane, DARPA's shapeshifting drone to explore distant worlds, solar sailing through space and NASA's developing technologies for a new space station and sustainable Moon bases.

My book takes you on an exciting journey of space discoveries, wonders, mysteries and explorations that you won't want to miss.

TABLE OF CONTENTS

AUTHOR'S BIOGRAPHY

Ed Kane created and serves as Executive Producer of CEO Global Foresight, a national news program on PBS focused on breakthrough innovations. He is the author of 23 books on the latest innovations across industries. Ed is a science graduate of the University of Pennsylvania.

1. PLANET X – THE BIG PERTURBER IN SPACE

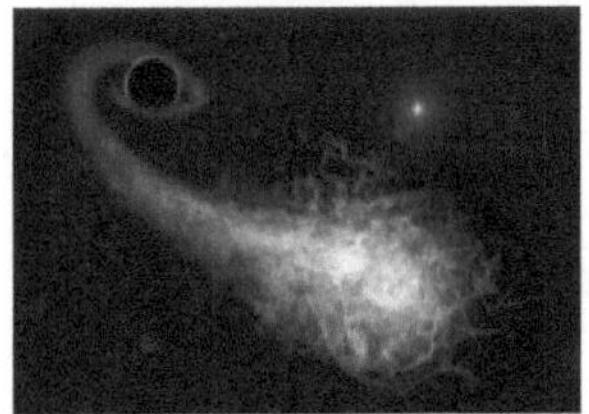
Source: NASA

Celestial Mystery on Edge of Solar System

The space phenomena on the outer edges of the solar system is known as Planet X, Planet 9 and Planet Next. And, global astronomers are baffled by it. They are not even sure that it's a planet. Some scientists speculate it's a grapefruit-sized extremely condensed and powerful Black hole. Its gravitational pull is so strong that it's also called the Big Perturber. Among the global astronomy community, opinion is divided. Some believe it is the new Planet X that is 5 to 10 times more massive than the Earth. Others think it is a powerful, extremely packed Black Hole with all of its explosive energy and mass inside a ball the size of a grapefruit. Quite a difference of scientific space speculation!

Tech to Determine by Late 2023

A powerful new piece of technology. - the highly anticipated Vera C. Rubin Observatory - is being built in the Chilean Andes. It is a massively powerful telescope that will go on line by the end of 2022. It has highly advanced technology that will be able to spot potential black holes, investigate dark energy and dark matter, find and track dangerous asteroids and study the Milky Way's formation and evolution. Several top astronomers at Harvard University are leading the exploration of Planet X. With the technology being deployed by the Rubin Observatory telescope, they expect to determine by late 2023 if the mysterious Planet X/9/Next/Big Perturber is actually a planet or an extremely powerful Black hole.

2. OBJECTS IN SPACE NEVER SEEN BEFORE

Source: Deep Space Image

Space: What's Up There?
Four mystery objects have been spotted by astronomers in deep space. The global team of astronomers say they've never seen anything like these objects before. Astronomers from the Royal College of Canada and Queens University, along with a team in Australia, discovered the objects using massive radio telescopes. They describe them as "edge-brightened discs", two are relatively close together and space objects "never probed before".

Deep Space Mystery
This discovery is one of the biggest mysteries in the universe. Astronomers are asking: What Are They? According to the team that discovered them: "We speculate they may be spherical shock waves from an extra-galactic event or a remnant from a radio galaxy viewed end-on". Whatever they are, they

EDWARD KANE

have the space world abuzz.

3. UK GOES INTO ORBIT

Source: OneWeb

OneWeb – Internet for Everyone, Everywhere
The United Kingdom and an Indian telecom giant are going into the satellite business together. The British government and Bharti Enterprises have bought OneWeb satellite company for $1 billion. The company was founded by US entrepreneur Greg Wyler with the purpose of providing internet everywhere to everyone with 648 low Earth satellites in orbit.

Rolling the Space Dice
This is a new lease on life for OneWeb and a gamble with potentially big payouts for the UK and Bharti. OneWeb was purchased out of bankruptcy by the UK and Bharti in July 2020. OneWeb filed for backruptcy late in March 2020 after its biggest financial backer SoftBank refused to provide any more funding. At the time of the filing, OneWeb had 74 satellites in orbit.

UK Mission and Goals
The UK is looking to add positioning technology to new satel-

lites in order to beef up the American GPS system it uses. It's also replacing bandwidth it is losing with the EU's Galileo network because of Brexit. There are estimates that the UK and Bharti will have to invest at least an additional $1 billion to bring OneWeb up to its full potential of providing continuous global internet service.

Crowded Space(s)

OneWeb is far from alone in the endeavor to provide universal internet through satellite constellations. SpaceX and entrepreneur Elon Musk are building a rival constellation network Starlink, now composed of about 600 satellites. When the number of OneWeb satellites hits 200, it may be able to go commercial. It's expected to first provide coverage at the North and South Poles for prospective military and gas and oil exploration customers. OneWeb is dual headquartered in the US and UK and needs to secure regulatory approvals for the purchase.

4. WORLD'S FIRST MARS HELICOPTER

Sourcee: NASA Ingenuity

Paired with Rover Perseverance

The world's first Mars helicopter - NASA's Ingenuity - will start operations in Mars' Jezero Crater in February 2021. It will be ferried there by NASA's Perseverance Rover. This is the first ever helicopter that will be flying in another world. It was developed by NASA along with Lockheed Martin and it is small, weighing just 4 pounds. The duo are expected to be launched from Cape Canaveral, FL aboard a rocket to Mars by late 2020. It's a great new mission of rocket science.

Delicate Maneuverings

Once established on Mars, the Perseverance-Ingenuity travel maneuvers will be delicate. Ingenuity will be attached horizontally under Perseverance for a hoped for February 18, 2021 landing at Jezero Crater on Mars. The landing clearance is tight. For 2 months, they'll stay together and hunt for a flat, unobstructed space to do test flying operations, at which time Perseverance will move football fields away.

Just the Right Spot

The decision-making is controlled by NASA experts on Earth. The right space for takeoff needs to be 33' x 33'. NASA operators will check all systems before Ingenuity flies. The process of unlocking the helicopter & making it completely vertical involves spectacular technology that even includes pyrotechnic fires releasing the legs. The helicopter will charge its batteries with solar panels.

30 Martian Days

Ingenuity is designed to work for 30 sols or Martian days to determine how feasible Martian helicopter flying will be for future missions. Ingenuity carries no instruments. It is providing the basis for the next generation of space helicopters for scouting difficult terrain, exploring deep craters and caves and even possibly carrying astronauts. As for Perseverance, the rover's main mission is searching for signs of ancient life on Mars and bringing samples back to Earth.

5. Center of the Solar System

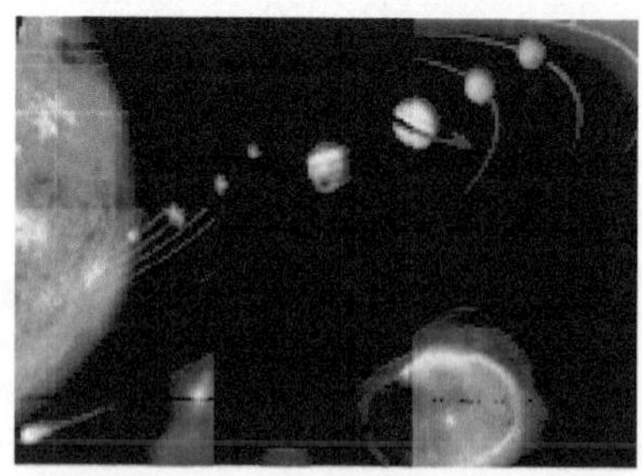

Source: Stock image of Solar System

International Astronomers Nail It

An international team of astronomers have narrowed down the center of the Earth's Solar System to within 328 feet. It is not dead center in the Sun as had been commonly thought. Instead, it's located just outside the surface of the Sun. The astronomers pinpointed the surprising location by monitoring the emissions of pulsars.

Pulsar Monitoring

This finding was not an easy task. There are a lot of gravitational forces influencing the calculation. The team used specially designed software to monitor pulsars. Pulsars are highly magnetized dead stars that emit electromagnetic radiation. Through the North American Observatory for Gravitations Waves, they were able to observe regular pulses from the pulsars and gravitational waves. Using that data, they were able to locate the center of the Solar System.

6. EU ACCELERATES SPACE RACE

Source: ESA Satellite

New Tech, New Satellites, New Innovations
The European Union is accelerating the investment of more money into its space explorations, satellite communications and rocket launches. Europe and the European Space Agency (ESA) have been very successful in space. They want to increase their efforts and keep pace particularly with the US and Chinese space programs. This, according to European Commissioner Thierry Breton in the summer of 2020.

EU Space Innovations
For the first time, the EU budget will be used to develop new technologies to launch rockets, including reusable rockets like SpaceX has developed. Breton has several novel ideas to spur space innovation:

- $1 billion euro Space Fund to support space startups
- $16 billion euros targeted just for space in the next EU

budget

- Competition to provide free access to launches and satellites for startups.

Bold Space Plan

This is a bold and innovative space plan for the EU. It includes rolling out a new generation of satellites in 2027 that interact with each other and provide more precise signals. It also includes the development of a new Space Traffic Management system to avoid collisions with the increasing number of satellites in space.

7. SUPER EARTHS DISCOVERED

Source: HARP Gliese 887 & Exoplanets

2 Exoplanets That Could Support Life

A team of German scientists has discovered two, Super-Earths just 11 light years away from Earth. The astronomers spotted them near the red dwarf star Gliese 887. They are outside of our solar system and they could sustain life.

Earth Like

University of Gottingen astronomer Sandra Saffers reported in the journal Science that the 2 exoplanets provide the best possibilities for the search for life outside of our solar system. In their report, the astronomers said they believe the exoplanets could contain liquid water and have a rocky surface like the Earth. They also have calculated that the surface temperature on these exoplanets at 158 degrees (F) or 70 degrees (C).

Gliese 887b & Gliese 887c

The newly discovered Super-Earths have been named Gliese

887b and Gliese 887c. NASA says more than 4,000 exoplanets have been discovered in space. Exoplanets have a mass larger than the Earth but smaller than the ice planets Neptune and Uranus. The German astronomers made their discovery using Chile's European Southern Observatory's High Accuracy Radial Velocity Planet Searcher (HARP) spectrograph.

8. VIRGIN GALACTIC, NASA AND ISS

Source: NASA

Virgin Galactic's Share Price Soars
Sir Richard Branson's Virgin Galactic and NASA are jointly developing a program for private missions to the International Space Station (ISS). Likely candidates for the private missions include scientific researchers, wealthy private citizens and space tourists. Virgin Galactic, like Elon Musk's SpaceX, is building a growing portfolio of business relationships with NASA.

Virgin Galactic's Role
Virgin Galactic will identify entities interested in buying private missions to the ISS. The cost is extremely pricey. A ticket to and from the ISS costs an estimated $58 million plus $35,000 per night onboard the ISS. The company will provide members of the private missions with astronaut training programs, transportation and necessary resources both while in orbit and on the ground as part of the preparations for the mission.

NASA's Strategy of Working with Private Companies

NASA is increasingly working with companies like SpaceX and Virgin Galactic as it prepares for its manned mission to the Moon in 2024. NASA's strategy is to reduce costs through the partnerships. In late May 2020 SpaceX successfully transported two US astronauts to the ISS. That was the first launch of US astronauts from US soil in more than a decade. NASA will debut a new space launch system for the journey to the Moon in 2021. NASA has big plans for a long-term presence on the Moon and a manned mission to Mars during the 2020's...all of which will be accomplished with strategic business partnerships.

9. STUNNING STAR FIREWORKS

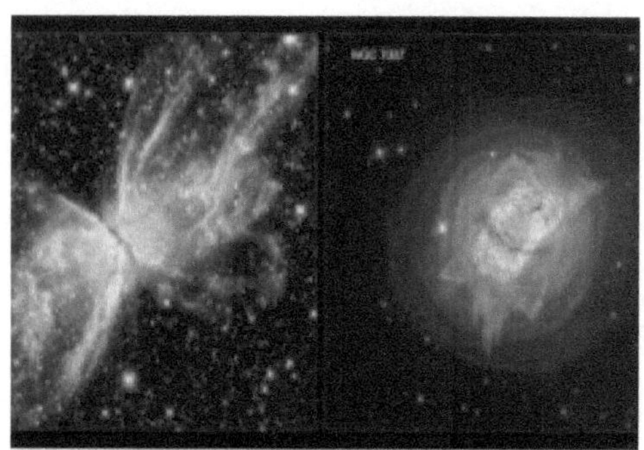

Source: Hubble Space Telescope

"Two Stars Going Haywire"

The NASA/ESA Hubble Space Telescope has captured new images of two nearby stars putting on a spectacular fireworks display in outer space. They are two planetary nebulae named the Butterfly nebula and NGC 7027. Astronomers say they are highly unusual, among the dustiest planetary nebulae ever discovered and contain unusually large amounts of gas.

Nuclear Fusion

Stars are engines of nuclear fusion. They live peaceful lives for billions of years. But near the end of their lives the fireworks start with erupting jets of hot gas from them exploding into the universe. Astronomers say the complexity and rapid changes in the jets of gas shooting off these two stars is unprecedented. The new Hubble images demonstrate that the two stars are splitting themselves apart in a very short timeframe.

Unprecedented Access

Hubble is providing astronomers with unprecedented access to this celestial event. They believe that NGV 7027 has been shooting off jets of its mass for centuries. But, according to lead researcher Joel Kastner: "Something recently went haywire in the center, producing a new cloverleaf pattern with bullets of material shooting out in specific directions." A team of astronomers led by Kastner is trying to pinpoint the cause of this spectacular, double star event.

10. SPACEX CREW DRAGON

Source: NASA - Crew Dragon Capsule

Many Historic Firsts on Earth and Space
In late May 2020, Elon Musk's SpaceX made history by carrying two US Astronauts to the International Space Station (ISS). This was the first time in nearly a decade that a rocket launched carrying US astronauts into space orbit from US soil. It was also a first for SpaceX to carry humans - NASA astronauts - into space. And it was the first time that a private commercial vehicle took US astronauts into orbit. Experts say this could be the beginning of a new era of human spaceflight for the US.

Historic Launch Time
The launch took place on May 31, 2020 from NASA's Kennedy Space Center in Cape Canaveral, Florida. President Donald Trump and Vice President Mike Pence were there. Two veteran US astronauts Doug Hurley and Bob Behnken crewed the SpaceX Crew Dragon capsule that rocketed into space on a SpaceX Falcon 9 Rocket.

Into Orbit

The Falcon 9 Rocket released Crew Dragon into a low Earth orbit 12 minutes after takeoff. During the next 19 hours, the astronauts orbited the Earth in Crew Dragon. They did some manual flying of Crew Dragon including raising the altitude of its orbit to move toward and reach the International Space Station.

Autonomous Docking with ISS

Crew Dragon has an autonomous docking system and didn't need much human help to connect and deliver the astronauts to ISS. Using a series of sensors and cameras to fly toward the ISS, it latched on to the open docking port of the ISS. Crew Dragon has successfully launched to the ISS before without any humans onboard. SpaceX has been developing this spacecraft and, along with NASA, testing it for six years.

Crew Dragon & Solar Panels

Crew Dragon operates with solar panels that last about four months in space. So the US astronauts will work in the ISS for a few months. When NASA decides it's time to return home to Earth, Astronauts Hurley and Behnken will get back into Crew Dragon, plunge back into the Earth's atmosphere and four huge parachutes will lower them into the Atlantic Ocean off of Florida. Crew Dragon has many emergency scenario, fail-safe systems to help protect the crew should there be a significant failure during the journey. This is rocket science, so we're wishing them the best of luck! The Falcon 9 Rocket is scheduled to return to Earth and land on a drone ship in the Atlantic Ocean when this historic journey is Mission Accomplished.

11. A GALAXY THAT SHOULDN'T EXIST

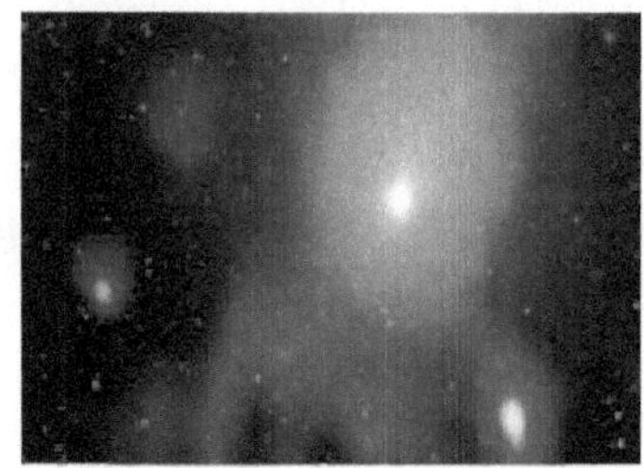

Source: ALMA

The Wolfe Disk

Astronomers have discovered a massive disk galaxy deep in the universe. It is almost as old as the universe itself at 12.5 billion years old. The universe is believed to be 13.8 billion years. It's called the Wolfe Disk and is yet another wonder in space just unveiled by astronomers, who are baffled by its very existence.

The Galaxy That Shouldn't Be

What's intriguing is that astronomers say the Wolfe Disk shouldn't exist and shouldn't be where it is in deep space. Those statements are based on astronomers' current understanding of galaxies. Their key question is: how did such a well formed rotating disk galaxy form at a time of chaos in space? A time when galaxies formed during violent collisions and chaotic mergers.

Created in a Different Way

The astronomers believe the Wolfe Disk was created and grew in a very different way. They theorize that it grew through the steady accretion of cold gas, which built its well-formed rotating disk. They're calling it the cold method of galaxy formation.

ALMA

Wolfe Disk is the most distant rotating disk galaxy ever observed. It was spotted by the Atacame group of massive telescopes, known as ALMA, in Chile. Wolfe Disk has a mass that is 70 billion times bigger than that of the Sun and rotates at 170 miles per second. It is quite a celestial find.

12. DISCOVERY OF CLOSEST BLACK HOLE TO EARTH

Source: European Space Agency

A Black Hole "In Our Neighborhood"

European astronomers have found the closest black hole to Earth ever discovered. It is so close that two stars circling it can be seen with the naked eye. The black hole is called the HR 6819 triple system. It was created by the death of a fleeting young star. Two remaining super-hot stars continue to orbit it.

Trillions of Miles Away From Earth

It is the closest black hole to Earth discovered thus far but it is still 1,000 light years away. Every light year equals 5.9 trillion miles. It was discovered by astronomers using the European Space Agency's massively powerful telescope at the European Southern Observatory in northern Chile. Lead astronomer Thomas Rivinius says the black hole "is in our neighborhood". He adds there may be 100 million to one billion of these relatively small but hugely dense objects in the Milky Way.

Black Holes Where Nothing Escapes

The difficulty of discovering black holes is nothing escapes from them. Even light doesn't escape so they are virtually invisible. In the case of this new discovery, astronomers found the black hole because of the unusual orbit of the two stars.

Dating Back 15 Million Years

This black hole was created 15 million years ago, according to the astronomers, when one of the three stars became too hot, too big, transformed to a supernova and then turned into a black hole in a violent process. The astronomers say the black hole is four to five times the mass of the Sun and its accompanying stars are three times hotter than the Sun.

13. NASA, VIRGIN GALACTIC AND SUPERSONIC TRAVEL

Source: NASA-Virgin Galatic

Flying Public at Supersonic Speeds

NASA is teaming up with Sir Richard Branson's Virgin Galactic space company to develop high speed technologies for supersonic air travel. The purpose is to make high Mach speed vehicles for the travelling public here on Earth to get from "Point-to-Point" cities on Earth in vastly reduced times. NASA and Branson finalized their collaboration by signing a Space Act Agreement.

Vision and Goal

Branson's vision for transporting the public at Mach speeds is a "global network of spaceports, transcontinental supersonic space flights and delivering passengers anywhere in the world in a couple of hours". To achieve this, the aircraft will actually be spacecraft.

Technological First

The envisioned SST would be a spacecraft that would launch into space from its departure point, then re-enter the Earth's atmosphere and arrive at the destination point for landing. The flight would happen in a fraction of the time a conventional plane would take. This spacecraft would be a technological first. The takeoff, space flight, re-entry, landing and reuse as a high speed civilian transportation system have never been achieved.

SSTs That Respect the Environment

Branson's deal with NASA brings his supersonic travel vision many steps closer to reality. And Branson says the supersonic spacecraft that he and NASA aim to develop will be environmentally responsible and sustainable. No timetable was offered as to when the new collaboration will start delivering new supersonic vehicles. But this concept brings space travel right down to Earth for global city to city travel.

14. HUMANITY'S FIRST LOOK AT BIRTH OF A PLANET

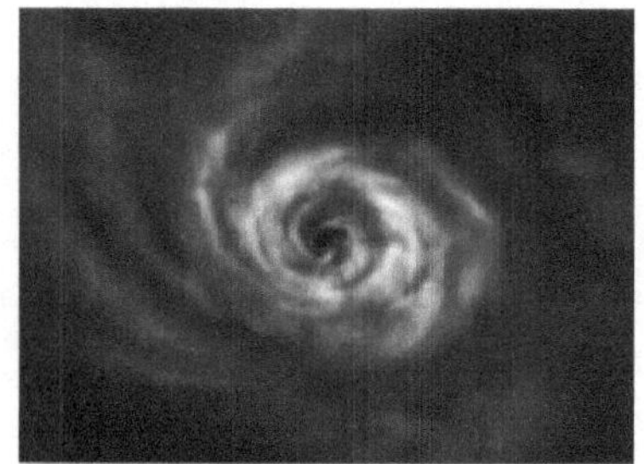

Source: European Space Agency

Creation of a Planet Deep in Space

Astronomers have discovered a new planet being born inside a swirling orange disc surrounding a distant young star. And, the team of international astronomers who discovered it says this is the first time that humanity has witnessed the birth of a planet. They add this discovery will help them to better understand how planets come to exist around stars.

Celestial Creation

As you can see in the picture, the new planet is forming in the very bright yellow twist in the inner region of the orange, swirling disc that surrounds the young star AV Aurigae. It was spotted by the European Space Agency's European Southern Observatory's Very Large Telescope in northern Chile.

520 Light Years Away

The birth of the new planet is taking place 520 light years from Earth. Lead astronomer Anthony Boccaletti of the Observatoire de Paris in France says the image shows cosmic matter at a gravitational tipping point collapsing and forming a new world. The international team of astronomers are from France, the US, Taiwan and Belgium. The image is the deepest observation of the AB Aurigae system to date.

15. SUN'S GRAND SOLAR MINIMUM

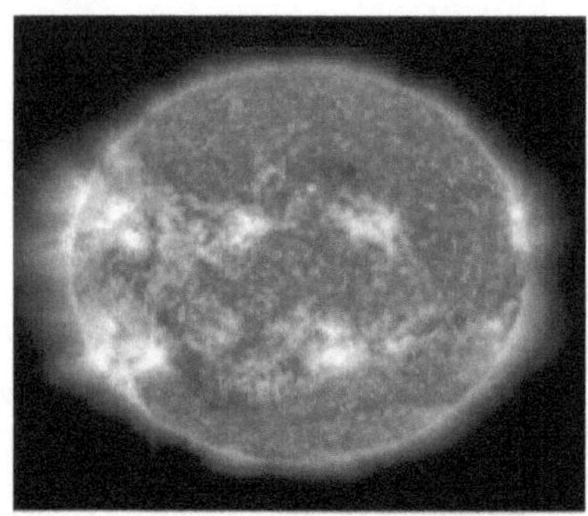

Source: NASA

Start of a New Ice Age?

NASA scientists say that the Sun is in a much less active phase, meaning far less solar flares, energy and sunspots. The phase is called a Grand Solar Minimum, which hasn't occurred since 1650 to 1715, when it triggered a Little Ice Age in the Northern Hemisphere with lower temperatures. The question is: Are we headed toward another Ice Age?

Climate Change

NASA scientists say we are not on the brink of a new Ice Age. And the new scientific basis of their conclusion is interesting. They say the warming of the Earth from fossil fuels and their greenhouse gas emissions is six times greater than even decades long cooling from the Grand Solar Minimum. In fact, NASA adds: "Even if the Grand Solar Minimum were to last a century, global temperatures would continue to warm." A shocking measure of the impact of global warming.

Solar Dynamics

The Sun is very dynamic and goes through 11 year cycles of peak and low solar activity, solar flares, energy and sunspots. The center of our Universe is ever changing.

16. 2020'S COMET SENSATIONS

Source: Google News

Cosmic Snowball

Comet SWAN is a cosmic sensation. Forbes magazine calls it The Comet of the Year. And, it was a visible phenomena to many in 2020. Comets aren't rare in our skies, but this one is. The reason: it was so exceptionally bright and beautiful to view. It put on some significant light shows especially in the Southern Hemisphere.

Comet Neowise

Comets are cosmic snowballs of frozen gas, rock and dust that orbit the Sun. Another comet sensation Neowise thrilled stargazers in mid-summer 2020. Its brilliant display of light won't be visible to Earth again for another 6,800 years. NASA says the last icy space wonderer to put on such a dazzling show was Hale-Bopp back in 1997. Neowise was so bright that it provided astronomers with a lot more and better data than most comets do. The scientists believe that the data they've gathered from multiple telescopes will provide key details about the composition and structure of the 3 mile wide comet Neowise.

17. MEGA SPACE TRAVEL TREND - RIDESHARING

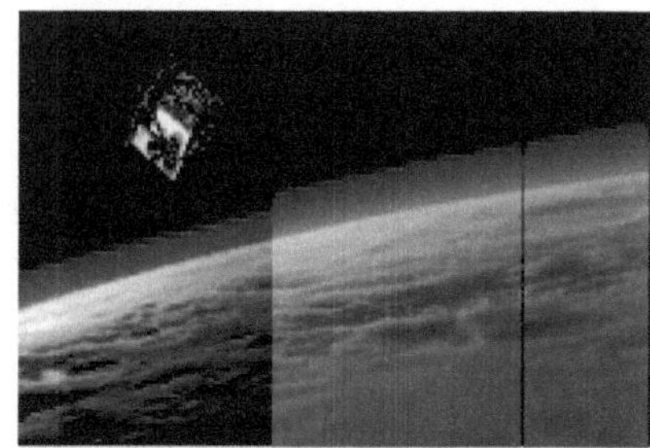

Source: Loft Orbit

Small Sats, Low Cost, Low Earth Orbit
Ridesharing is a new megatrend in Space. A SpaceX Falcon 9 rocket will fly on a ridesharing mission, late in 2021, carrying a Loft Orbiting satellite with a DARPA Blackjack Pit Boss mission system inside for a space demonstration. Loft buys space buses (satellites) and rents space. The US Defense Department's Advanced Research Projects Agency (DARPA) will be flying onboard the SpaceX rocket with many other customer payloads. DARPA's Blackjack programs aims to deploy low cost, small, low earth orbit satellites into operation for the US Military.

Space Experiment
DARPA's testing of Blackjack will commence in early 2021. The experiment is called Sagittarius A. It's designed to demonstrate any risks or problems before the main Blackjack satellites go

into production and launch in late 2021.

Pit Boss

Pit Boss is the software mission management system that will enable the satellites to autonomously acquire, process and provide information to the system users. Sagittarius A demonstration will test the flight computer and an optical imaging sensor that searches for targets in the ocean. The data goes into the satellite's flight computer, AI processes it and then redirects the satellite to start searching for more targets. This real, dry run in space is expected to usher in the deployment of Blackjack in 2022. It will also contribute to increased "Ridesharing" to space.

18. SUPER-SECRET USAF SPACE PLANE

Source: MIT Technology Review

Launched May 2020 and On Duty

The US Space Force & USAF's top secret mission of its unmanned, robotic, long duration, X-37B space plane is underway. It hurtled into space in May 2020 onboard a United Launch Alliance Atlas V rocket for what is now being called the US Space Force's OTV-6/USSF-7 mission. This is another secret mission for one of the most fascinating space planes in the universe.

Green, Solar Powered, Space Plane

The X-37B is an unpiloted, long distance, long duration flight spacecraft. It flies classified missions that can last months and even years for the US military. It is a green machine that's powered by a solar array that's the size of a pickup truck and embedded inside the space plane.

Tremendous Technology

The technology is incredible. There are 2, X-37B space planes built by Boeing. The winged aircraft can land themselves back on Earth, are reusable and look like a small NASA space shuttle. Plus, they are solar array powered. They have been on six missions since 2010 with 2,865 days in orbit.

Mission Experiments Onboard

The current mission for X-37B, like all the other missions, is classified and top secret. But it is known to contain several interesting experiments, that include a small Falcon Sat-8 satellite invented by the USAF Academy, NASA space environment experiments and a novel solar power beaming experiment from the Naval Research Laboratory.

19. NASA'S GROUND RULES FOR LIVING ON THE MOON

Source: Stock Moon image

NASA's Artemis Accords

NASA has released a set of basic principles designed to govern how humans will live and work on the Moon. They are called the Artemis Accords and include the main tenets of what is designed to become an international agreement for the exploration of the Moon. It has sparked international debate. Russia is angry that it's not part of the early negotiations. In fact, the Russian space agency Roscosmos has accused the Trump Administration of "planet snatching" in its efforts to drive formation of the Accords. China too has complained that it isn't part of the early negotiations.

Lunar Safety Zones and Mining Rights

The Accords are designed to establish safety zones around lunar bases to prevent what NASA calls "harmful interference". They would permit companies and nations to own the lunar resources that they mine. The Accords are part of NASA's strategy to win support from allies for its plans to build long term bases on the Moon as part of NASA's Artemis lunar mission.

International Space Law
At this point, the Trump Administration is bypassing the United Nations and attempting to craft bi-lateral and multi-lateral agreements on the Moon accords with allies. President Trump wants the Accords to establish under international law the rights of companies to mine the Moon and own those resources. Some Russian space experts have argued those lunar resources are the property of all humanity.

20. JOURNEY TO THE SUN: NASA AND ESA'S BLACKBIRD MISSION

Source: NASA Launch of Solar Orbiter

Unprecedented Solar Orbiter Probe

NASA and the European Space Agency have successfully launched the Solar Orbiter on a journey to the Sun, to take unprecedented views of its blazing solar poles. The Solar Orbiter took off on top of an Atlas 5 rocket. The Orbiter separated from the rocket within 53 minutes to go into space flight as planned and is now in communications with NASA. All of this happened in February 2020. This unique mission is going very well as it moves along on a very long journey to the Sun.

Mission Unprecedented

This US and European mission is unprecedented. It's expected to help scientists understand how the Sun's massive amount of energy affects humans in space and how it impacts all of us on the Earth. The mission will provide unique images, views and data of the sun's blazing pole regions, which are critical sources of data.

Ten Year Journey

The orbiter is equipped with awesome equipment including an array of solar panels and antennas. The journey to the Sum will take ten years. The orbiter will position at 26 million miles from the Sun, which is 95% of the distance between the Earth and the Sun. It will map the Sun's poles which have a concentrated source of solar winds. Those winds are high impact. They penetrate our atmosphere and even can impact satellites.

Main Goal

NASA and ESA scientists want to determine how the Sun creates and controls the heliosphere, which is the massive bubble of protection that surrounds the solar system. They want to know why the bubble changes over time. NASA and ESA scientists believe the answer may be found in the Sun's poles. Those views will first be available from the probe in 2025. Exciting new science for all of us to observe from Earth!

Key Goals to Get This Off the Ground

The key to this mission, that was first proposed in 1999, was to develop a thermal protection system that can withstand the intense heat of the sun. The Orbiter was built in Europe. It's a great example of US-European space exploration. The cost of the Solar Orbiter hurtling into space is $1.5 billion.

21. VIRGIN GALACTIC'S AIR TRAVEL REVOLUTION

Source: Virgin Galactic

LA to Tokyo in 2 Hours

Billionaire entrepreneur Sir Richard Branson is a brilliant innovator. And he has beyond supersonic plans for his space travel company Virgin Galactic. Sometime in 2021, he expects to have flown the first commercial tourism flight into space. Nearly 8,000 tourists have signed up for a space flight at a cost of $250,000 each. But Branson business plans go way beyond space tourism. He wants to revolutionize commercial air travel by using his space vehicles for greatly accelerated air travel here on Earth.

Beyond Supersonic Speed - Spaceline Flying

Virgin Galactic says it is the only company that can deliver flying at greater than supersonic speed in a winged aircraft. It calls the high speed flying "Spaceline" flying. The company says its spaceship Unity and two other space vehicles under construction can be used to vastly accelerate long distance fly times, for instance from New York to Paris. The company pinpoints the market potential for this type of flying at $900 billion. Just capturing the premier business flyer portion of the market would be worth billions.

High Speed Global Mobility

Virgin Galactic's business plans for growth are strategic and aggressive. Besides space tourism, it plans on offering extremely high speed air travel that could cut the fly time between LA and Tokyo to two hours. As part of that, Virgin Galactic will develop what it calls "high speed mobility vehicles". Also it's pursuing commercial and government users of its proprietary technologies.

22. SNOWMAN STAR BORN BY TWO STARS MERGING

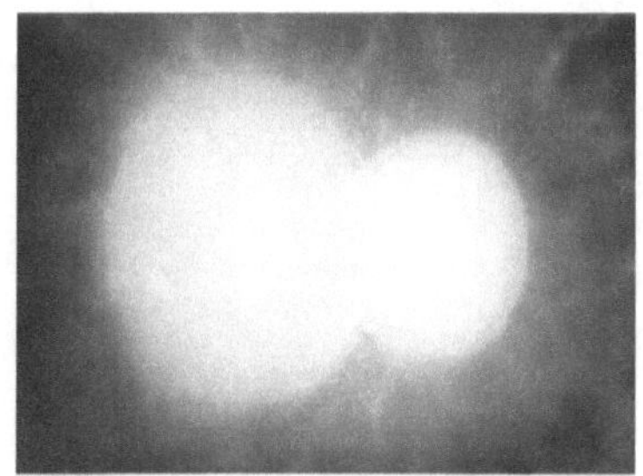

Source: Warwick University, UK

Unique Shape and Unique Atmosphere Composition

Nothing like this new star has ever been discovered in space before. Warwick University scientists have spotted a huge, snowman-shaped star. It's a new star born by the merger of two "white dwarf" stars. They've named the star WDJ0551+4135. The star's atmosphere is unique and unusually rich in carbon. Everything about this star has the potential of leading to new scientific discoveries about the universe and the birth and death of stars.

Bigger Than the Sun

The new Snowman star is more massive than the Sun. Another unusual aspect of the birth of this star is that the merger of the two "white Dwarf" stars didn't result in an explosion and powerful supernova. The Warwick University scientists say what is particularly exciting for their research is that the stars merged rather than exploded, as is usual.

How a Star is Born

The scientists believe that measuring the properties of the failed supernova will tell them a lot about the "pathways to thermonuclear self annihilation" of stars. A "white dwarf" is what stars like the Sun become after using up all of their nuclear fuel. In this case, a new star was born by the successful merger of two starts.

23. NASA SPACE TOURIST DESTINATION FOR YOU

Source: Axiom

NASA Partners with Axiom

Space exploration company Axiom of Houston, Texas is partnering with NASA for a big launch of space tourism. Tourists will be able to visit the International Space Station (ISS) as early as 2024. Axiom is a private US space station manufacturer that is building, what it hopes will be, the world's first commercial, international space station. CEO Michael Suffredini is the former program manager of the ISS for NASA. This is a big business alliance, with tremendous expertise behind it, for launching tourists into space.

Advanced Luxury Space "Hotel" Priced as High as $35,000

The Axiom space vehicle is an impressive piece of technology.

First of all, it will be led by a trained astronaut, have a crew and also research labs on board. It has the biggest space observation window ever built into a space capsule and the state-of-the art interior was designed by the top French-based hotel and yacht designer Philippe Starck. This takes space tourism to a whole new level.

Space Nest

The interior is designed like a "nest" according to Starck with hundreds of nanoLEDs that change color and embedded touch-screens and handles. It will also have Wifi. Starck said the space provided him as a designer and the future occupants what he calls "multidimensional freedom" as it is zero-gravity with no horizontal, vertical or diagonal limitations.

Ticket Takes Money & Fitness

The Axium-NASA module is designed for fully trained astro-nauts whose nations aren't members of the ISS and for private citizens. If you can pay the price, you also need to be very fit to get a seat. Travelers to space must clear a physical and then 15 weeks of very rigorous training that includes jet flights, ex-treme environment endurance and suborbital space flights. If you clear, you get one ticket to ride.

24. BEYOND THE BIG BANG

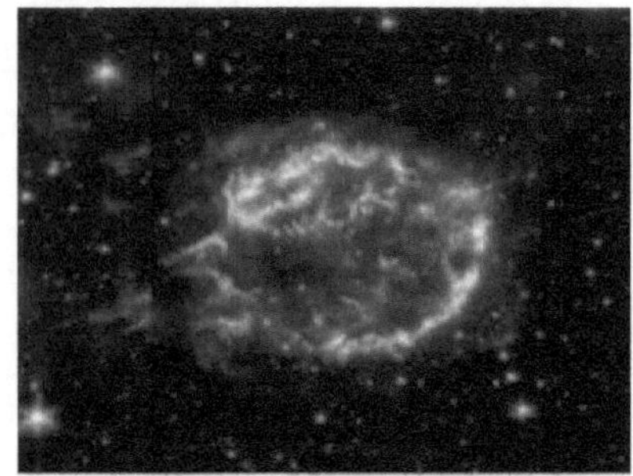

Source: NASA

Huge Explosion in Universe From Light Years Ago Detected
Astronomers using data from the most advanced NASA, European Space Agency, Indian and Australian telescopes have discovered the biggest explosion that the universe has ever experienced since the Big Bang. It happened in a galaxy 390 million light years away from the Earth. And incredibly it took place in slow motion over hundreds of millions of years. The science to discover this phenomenom is remarkable and the details emerging on events going back into space from such a distance and time are amazing, innovative scientific discoveries.

Supermassive Black Hole in the Ophiuchus Galaxy Cluster
The global team of astronomers say the explosion they detected released five times more energy than ever detected before. The explosion came from a supermassive black hole in a galaxy called the Ophiuchus Galaxy Cluster nearly 400 million

light years from the Earth.   The power of the explosion was enough to punch a hole into the gas surrounding the black hole that is big enough to contain 15 contiguous Milky Way galaxies. This is an amazing scientific discovery through breakthrough telescopic innovation. What it means to the formation of the Universe and Earth as we know it is the next chapter of the history of the Universe.

25. NASA DISCOVERS MARS IS A DYNAMIC PLANET

Source: NASA

Hundreds of Marsquakes in the Past Year

NASA's robotic lander InSight delivered some big discovery and innovation dividends. InSight detailed, for the first time, that the Red Planet Mars experiences quakes like the Earth and the Moon. NASA has called the shake, rattle and rolls on Mars "Marsquakes". Incredibly, InSight has detected 450 seismic events on Mars in the past year as it has explored Mars on the ground. 20 of the quakes were significant and in the 3 to 4 point magnitude range.

NASA First

According to NASA, this is the first time that they've established that Mars is a seismically active planet. InSight's observations will help scientists better understand how rocky planets like

Mars and the Earth form and evolve. And the lander's sensors also detected significant winds swirling around Mars. In fact, there are thousands of passing whirlwinds whipping around Mars.

Mars Quakes

Compared to quakes on the Earth and the Moon, the Mars quakes are relatively small. Mars is more seismically active than the Moon but less so than the Earth. The Marsquakes provide NASA scientists with new data and information on the interior core of Mars. NASA says the quakes on Mars are caused by the long-term cooling of the planet that makes it contract and facture.

Tech Wonder

InSight is a robotic technological wonder. The lander is equipped with heat flow probes to take the planet's temperature, sensors to gauge wind and air pressure, seismometers to detect quakes and a magnetometer to detect magnetism.

26. STARDUST DISCOVERY OLDEST ON EARTH

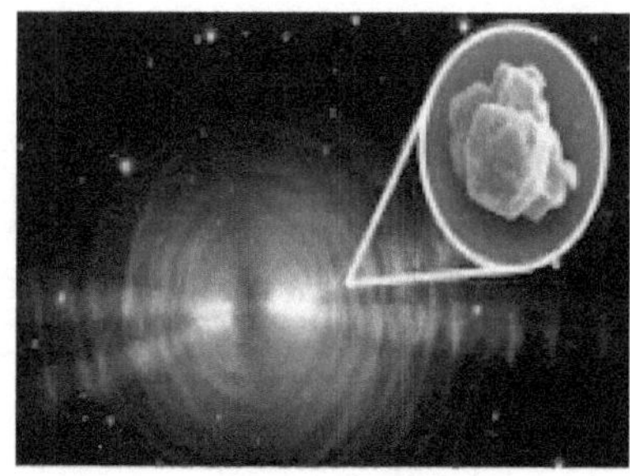

Source: NASA Meteorite

Stardust Material That's Older than the Sun

This is an extraordinary scientific discovery announced by a team of astronomers led by University of Chicago Professor Dr. Philipp Heck. In a meteorite that fell from space to Australia 50 years ago, there is awesome stardust dating back 5 to 7 billion years. This stardust predates the Sun. Heck and his team analyzed the contents of the meteorite. They found it to be the oldest, solid material ever found on Earth. Their work has been published in the prestigious Proceedings of the National Academy of Sciences.

What This Discovery Opens Innovative Views On

The scientists say this stardust will tell us how parent stars were formed in our galaxy. It will tell us about the origins of the oxygen that we breathe and provide the opportunity to trace materials back before the creation of the Sun. The stardust is the oldest material discovered to reach the Earth. The material is older than our solar system. The Earth is estimated to be 4.5 billion years old. The Sun is 4.6 billion years old. The newly found and analyzed stardust is 5 to 7 billion years old. This discovery does give a whole new perspective on our brilliant and beautiful universe!

27. SATURN'S TITAN MOON AND LIFE

Source: NASA Jet Propulsion Lab

Moonscapes, Science and Life

NASA scientists have mapped Saturn's exotic, largest Moon Titan. Apparently, Titan has quite a story to tell about the possibilities of life. NASA has unveiled the first global geological map of Titan. It shows a strange and exotic world that NASA scientists say is a strong candidate for the search for life beyond Earth.

Incredibly Interesting Terrain

The map shows dunes of frozen organic material, vast stretches of plains, and lakes and seas filled with liquid methane...all of which may harbor forms of life beyond Earth. The map is based on radar, infrared and data from NASA's Cassini spacecraft that studied Saturn and its moons from 2004 to 2017. The data and images now emerging are fascinating.

What's Next

Clearly, Titan contains organic material that is the ingredient critical to form life. What's next is NASA is going to launch its Dragonfly Mission sometime in the 2020's to reach Titan by 2034 with a multi-rotor drone to explore this exciting new Moon in space.

28. SPACEX UNVEILS MARS VEHICLE

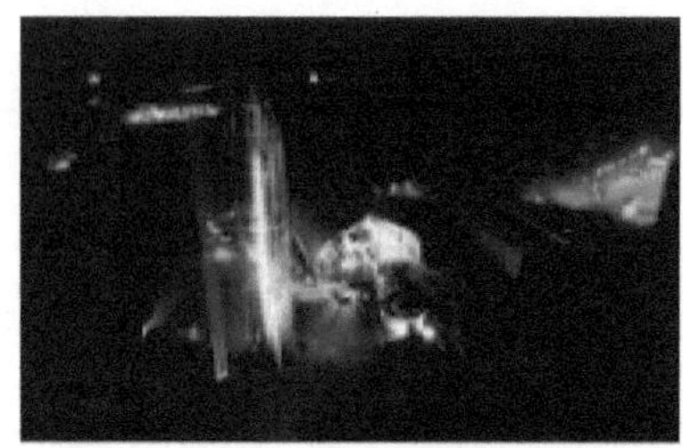

Source: SpaceX Starship

CEO Elon Musk on to Mars

SpaceX CEO Elon Musk unveiled his silver bullet: a spacecraft to take you to Mars and beyond. It's the new Starship Mk1 prototype, unveiled at SpaceX's South Texas test site in Boca Chico, TX. Starship is a massive, reusable launch system. When finalized, the vehicle will have a 387 foot Starship Super Heavy stack and be powered by six Raptor engines. The spacecraft has room for 37 Raptor engines, depending on the mission needs and distance.

Musk, a Visionary Thinker

Starship will be capable of taking up to 100 people back and forth to space, including the Moon, Mars and other space destinations. Musk says space travel has to be done like air travel. He believes that is the first breakthrough needed to make humans a space-faring, multi-planet civilization. He adds making space travel like air travel is "the fastest path to a self-sustaining city on Mars", which is his ultimate goal and destination.

Passengers in the 2020's
Musk says the Starship vehicle system could soar into space and back to Earth three times a day. And, if all goes well, people could start flying as passengers on the vehicle sometime during the 2020's

29. NASA'S SHAPESHIFTER EXPLORING NEW DISTANT WORLDS

Source: NASA/CalTech Concept

Source: NASA

Drone of Drones

NASA's Jet Propulsion Lab at Caltech has introduced the drone of drones. It separates into two different units: two shapes to explore new distant worlds and new habitats. It's a prototype and is made of several small, quadcopter drones called cobots. The cobots have propellers and fly independently and the system reassembles to roll on the ground. The future for the concept is much bigger. The Shapeshifter would be made up of a number of small robots that can easily self-assemble into larger robots and disassemble as the mission requires, particularly in space. The mini-robots will be able to fly, roll, float and swim and then morph into a single machine.

Morphing Robots

On the ground, the cobots come together to form a roller-wheel like drone to explore the ground in places in space like Titan, where there is very limited information about the surface. The unpredictability of the surface makes versatility and shapeshifting essential in the drone. In the future, the plan is to make the cobots work together as a team of twelve to explore caves, underwater areas and various types of terrain, including in outer space. The ultimate morphing robotic team would be carried aboard a mothership lander that would house their energy source and scientific instrumentation for testing and analysis.

Dragonfly 2026

This extraordinary technology will take quite a few years to fully develop. But a deployment target date for it is 2026 when NASA's Dragonfly drone takes off for Titan with possibly the Shapeshifter onboard. The Shapeshifter is a transformational vehicle to explore, treacherous, distant worlds.

30. HOW TO GENERATE LIGHT AT NIGHT

Source: UCLA

Renewable and Complements Solar Power
This new, inexpensive thermoelectric device may be transformative in energy generation. It harvests the coldness of space during the night to generate electricity - enough right now to power a LED light at nighttime, but the inventors say it's very scalable. The device is a significant new innovation from engineers at UCLA and Stanford University. The gadget works at night when solar systems don't. The inventors say it's a new approach to power generation when power at night is needed. It complements solar power that doesn't work at night, giving a 24/7 approach to green, renewable energy.

Phenomenon Like Frost Formation
The device takes advantage of radiative cooling, the process by

which frost forms on grass during above freezing temperatures at night. The sky facing surface of the technology passes heat to the atmosphere as thermal radiation. It loses some heat to space and reaches a temperature cooler than the surrounding air. That temperature differential produces renewable energy at night, when lighting demands are peak.

Scalable Tech for Global Use

According to the UCLA and Stanford engineers, their invention is highly scalable. The radiative cooling device essentially consists of an aluminum disk coated with paint and all the other components are readily available for purchase off the shelf. This is important innovation to watch for because of its practicality and scalability for worldwide use to supplement solar energy at night.

31. PLANET RAINING DIAMONDS – NASA DISCOVERY

Source: MIT

Discovery Unveiled

NASA scientists have discovered that Saturn's mysterious atmosphere is raining diamonds down onto the planet. This discovery is extraordinary and yet another example of the vast mineral deposits in space that commercial space ventures are targeting.

Saturn Rings

While studying how and when Saturn's rings of ice and dust emerged, the NASA scientists stumbled upon a surprising feature of Saturn's atmosphere. The atmosphere is composed of high amounts of sulfur, along with hydrogen and helium. The atmosphere is extremely harsh and unforgiving. So much so it rains diamonds onto the planet. NASA's Cassini probe made the shocking discovery when taking a closer look at Saturn's weather in 2013. The discovery was unveiled in a recent BBC documentary.

From Soot to Diamonds

Saturn is the sixth planet from the Sun and the second largest in the Solar System. According to scientist Dr. Brian Cox, Saturn's atmosphere is brutal. There are huge clouds filled with water and lightning that is 10,000 times stronger than on Earth. The combination changes methane gas in the atmosphere into huge clouds of soot. The pressure is so great the chunks of soot transform into diamonds, which rain down on the planet as liquid because of the intensity of Saturn's environment.

32. SUPERNOVA FOUND IN ANTARCTIC SNOW

Source: Stock Image Antarctic

Interstellar News

Australian scientists have discovered large amounts of stardust in Antarctic snow, likely from the explosion of a supernova that had been close to the Sun. The explosion rained down particles of a unique iron isotope discovered in melted snow from the South Pole. The discovery of the rare isotope, iron-60, which is not native to Earth, was made by a scientific research team from Australian National University.

South Pole Supernova

The researchers have ruled out any chance that the isotope was created by human activity. They say the only explanation is that it was created by an interstellar rock hitting the Earth. The team melted more than 1000 pounds of snow and analyzed the contents of the melt, which contained large amounts of isotope

iron-60. Their next focus is to determine the age of the isotope and to dig deeper for more samples in the icepack. Their
remarkable findings were published in the journal Physical Review Letters.

33. UNPRECEDENTED COSMIC EVENT

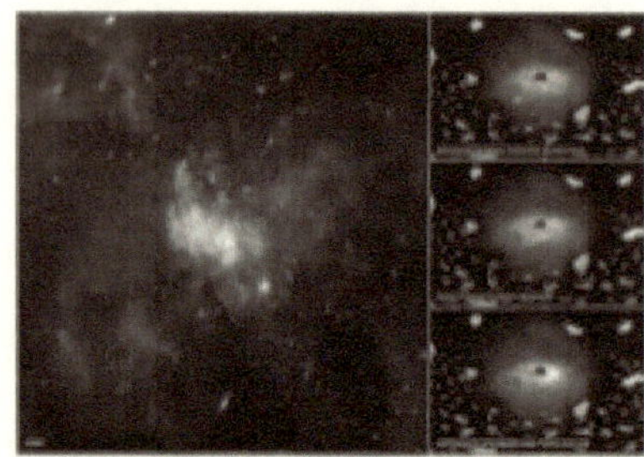

Source: NASA Sagittarius A

Sagittarius A: Black Hole Light Show

The black hole Sagittarius A (Sgr A) is the Earth's closest black hole and is right in the middle of the Milky Way. Using the Keck II Telescope in Hawaii, astronomers watched the massive black hole erupt and spew a huge burst of infrared radiation. They call the event "unprecedented" and they can't explain exactly what caused it.

Off the Charts

The astronomers say the black hole reached much brighter "flux levels" than ever observed before. According to NASA, SgrA is 26,000 light years from Earth. The astronomers disclosed it in the journal Astrophysical Journal Letters. According to the astronomers, black holes are always variable but this one was off the charts of historic data including from the Keck II Telescope.

Theories

There are two theories as to what caused the huge burst of light. A star passing by could have changed the gas flow around the black hole. Or it could have been a flash from a passing gas cloud.

34. MARTIAN ROCK TELLS BIG STORY

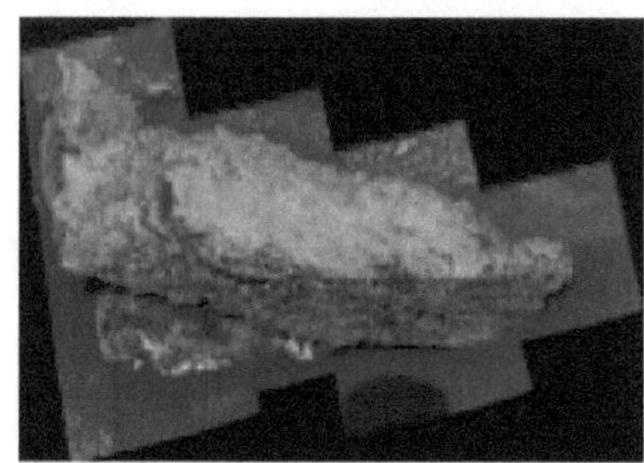

Source: NASA

Shows Presence of Water, Raises Questions of Life on Mars
NASA's Curiosity rover has been exploring the planet Mars for the past seven years. It found a rock that contains some of the secrets to what Mars was like when it was a lot more wet. The discovery is so important that the rock has been nicknamed "Strathdon".

Strathdon
The rock was found in the Gale Crater or the "clay bearing unit". The rock is composed of layers of sediment that are unique and wavy. NASA scientists say this shows evidence of flowing water and blowing wind. They add the history of water on Mars is much more complicated than they had previously thought. They now do not believe that Mars went from wet to dry overnight.

Signs of Life on Mars
The question still remains: with the presence of water on a

warmer Mars with a thicker atmosphere, could that have sustained more than microbial life? NASA is searching for that answer.

35. STEAM-POWERED TWIN SPACECRAFTS

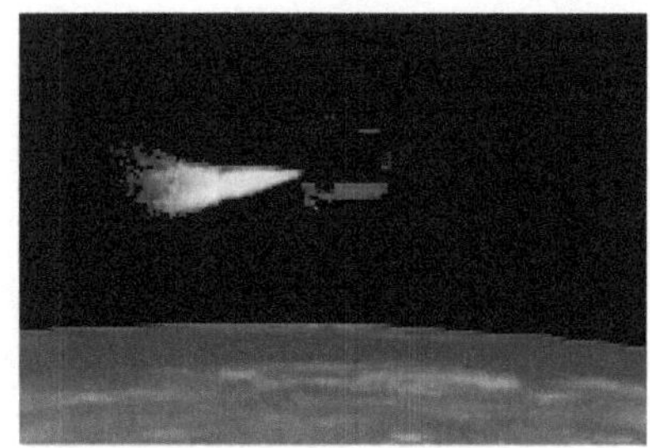

Source: NASA Cubesat

High Theater in Low Earth Orbit

Two cubesats (small satellites) powered by steam have successfully performed a complicated and coordinated maneuver in space. With the help of NASA on the ground, one of the spacecraft commanded the other to close the 5.5 mile gap between them. It did so successfully.

Old Fashioned Steam Power

The twin spacecrafts' fuel tanks are filled with water. Thrusters convert the water into steam which propels the cubesats. NASA says the maneuvers demonstrate how small spacecraft can work together on future missions. NASA adds these spacecraft are designed to operate in swarms. These space teams could be deployed in the future for deep space exploration and work autonomously to explore new worlds. NASA plans on sending cubesats to the Moon in the 2020's.

36. STAR THAT'S OLDER THAN THE UNIVERSE

Source: NASA's Hubble Shot of the Star

Star Discovered That May Be Older Than Our Universe

It's a very old star discovered by astronomers that's a cosmic riddle of intergalactic proportions. Astronomers have found a star that they think may be older than the universe. Best estimates by astronomers is that the universe is 13.8 billion years old. But they have found a star close to earth that is estimated to be 14.5 billion years old, based on its very low metal content. The star is HD 140283, nicknamed the Methuselah Star. It's raising a lot of questions, fascination and wonder. Is that beautiful twinkling star older than time and the universe itself?

The Methuselah Star

The questions come from many including a top physicist at the UK's Royal Astronomical Society Dr. Robert Mathews. He calls it a riddle of cosmic proportions. How can the universe contain stars older than itself? This question sparks questions about how accurate calculations are about the age of the universe. Even NASA's estimate is imprecise: Methuselah could be 800 million years younger or older than 14.5 billion years. It is too early to tell but Dr. Mathews is looking into new research into gravitational waves as a possible explanation for the age differential. But, he adds the age of the universe question is back with a vengeance.

37. CHARTING THE MILKY WAY

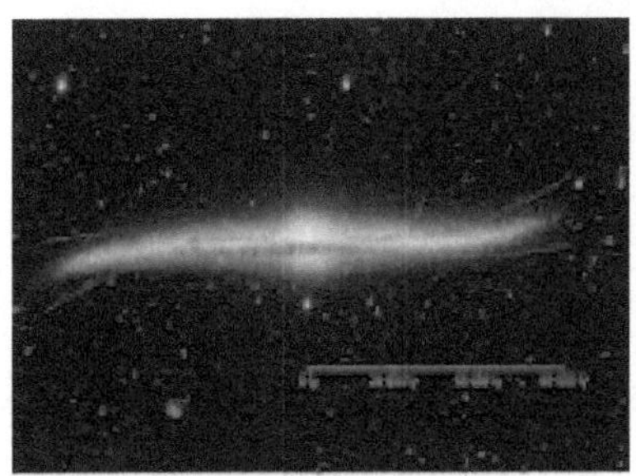

Source: University of Warsaw

Surprise Finding "Warped and Twisted", Not Flat
Polish astronomers from the University of Warsaw have created the most precise and largest map of the Milky Way to date. By tracking thousands of blinking stars in the galaxy, they've created a 3D scale map of the system. They discovered that it isn't flat as a pancake but twists and turns in shape much more than previously thought by astronomers.

Spiral Galaxy
The Milky Way is a spiral galaxy measuring 120,000 light years across. It's a mass of stars, gas and dust about 27,000 light years from the Earth. The Polish team found the distortions in the Milky Way are immense with some stars 60,000 light years away from the Milky Way's center. And, the galaxy's thickness is variable.

Potential Causes

The astronomers cited several potential causes for the variations including interaction with nearby galaxies, intergalactic gas and even dark matter. Their groundbreaking work was published in the journal Science.

38. SUPER EARTH FOUND IN SPACE

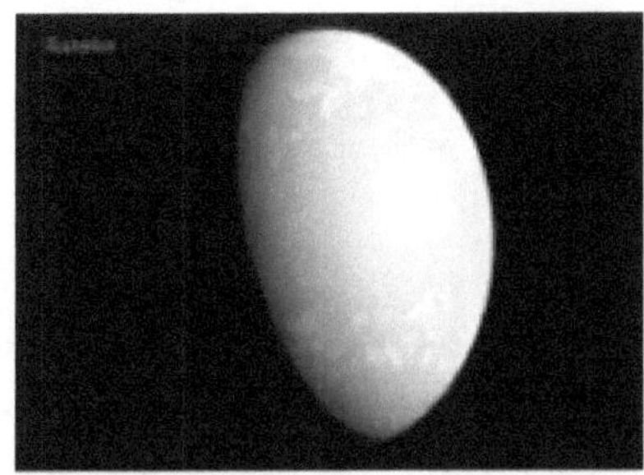

Source: NASA's Goddard Space Flight Center

May Have "Earth Like" Conditions

Astronomers have discovered what they're calling a Super Earth 31 light years away from our solar system. They say it is "potentially habitable". The planet is named GJ 357d. It is six times larger than the Earth and it orbits a sun that is much smaller than ours. An international team of scientists discovered it using powerful ground telescopes and following up on data first relayed by NASA's planet hunting satellite TESS. The astronomers believe that it could provide "Earth like conditions" for life.

Looking for Signs of Life

Highly advanced telescopes will, in the next several years, be going on line that will be able to pick out any signs of life on Super-Earth. Two telescopes going live in 2021 and 2025 should show whether the planet is rocky and if it has any oceans. If the planet's atmosphere is thick, it could support water on the surface and sustain life. If there is no atmosphere, the average temperature would be 64 degrees below zero, making it more glacial than habitable.

39. SOLAR SAILING THROUGH SPACE

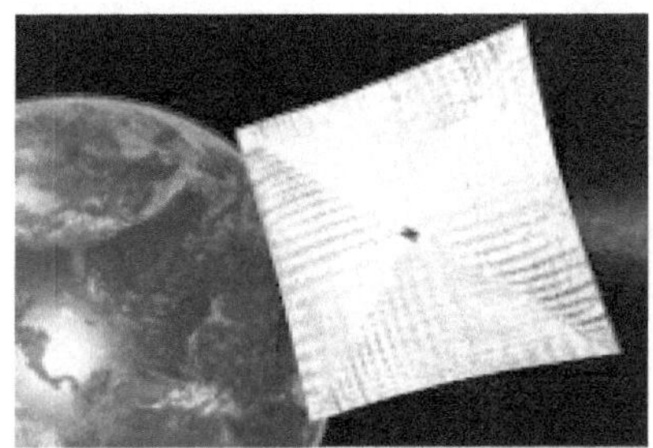

Source: The Planetary Society's LightSail in Orbit

Solar Powered Space Missions

This green energy technology has the potential to be disruptive innovation. It's a solar powered sail invented by the non-profit Planetary Society in California, led by the Science Guy Bill Nye. The LightSail 2 CubeSat spacecraft successfully deployed its solar sail in space. It's the first spacecraft to be propelled by sunlight alone. This is very important science because the use of solar sails to convert solar wind into thrust could save fuel and provide clean solar power for spacecrafts on long space missions.

Solar Sailing

The small spacecraft is the size of a toaster. It was hurled into space aboard a Space X Falcon Heavy rocket in the summer of 2019. Flight controllers on the ground in California fully deployed the solar sail from the satellite into space. The sail is the size of a boxing ring and the system is orbiting the Earth solely on the power of the sun. This system is designed to test solar sailing technology and works very well. It's in full orbit and has relayed pictures back to Earth.

Photon Power

The LightSail 2 is a square composed of four triangles and made of aluminized Mylar. When the sunlight's protons hit the Sail, the pressure creates solar wind and pushes the craft forward. LightSail has completed its original year long mission in space and has entered an "extended space mission" phase to optimize the solar sailing control software and the vehicle's space performance.

40. NASA'S 21 WHOLE NEW WORLDS

Source: NASA

Planet Hunter Extraordinary

NASA's planet hunter TESS spacecraft - Transiting Exoplanet Survey Satellite – recently marked its first year in space and the results are extraordinary. The spacecraft has discovered 21 new worlds outside of our solar system. It's also spotted 850 possible exoplanets that have to be confirmed and 6 supernovas. For NASA the results are way beyond all expectations.

Star Search

TESS is focusing on stars that are less than 300 light years away. It looks for dips in brightness which indicates that an object just passed across the star. The data will help NASA determine which exoplanets it wants to explore in-depth and where life might exist.

Historic Mission

This is the most comprehensive search for planets ever accomplished by mankind. TESS finished exploring the southern part of the sky and is examining the northern sky.

41. SUSTAINABLE LUNAR BASES

Source: Stock Image of Future Lunar Bases

Fascinating Research by the European Space Agency

Administrators of the European Space Agency (ESA) believe that the next step in space exploration is building lunar bases for astronauts. To do that a sustainable source of energy is required to sustain human life and power for rovers and landers. ESA scientists think that they have one, using the surface of the Moon. ESA scientists have created bricks composed of lunar regolith, which is the soil, dust and rocks on the surface of the Moon collected on past lunar missions. In testing, the bricks are able to store solar energy and generate it for use as power, heat and electricity. It's possible this innovation could provide future lunar bases a sustainable energy source.

Scaling Up the Technology

The ESA team is now scaling up the process and efficiency of their heat energy bricks from lunar regolith. They say that travelers to the Moon, with this new energy source, wouldn't have to take much with them from Earth. They also think it has the potential to enable very ambitious missions into space. As the world marked the 50th anniversary of the Apollo 11 historic mission to the Moon, the work being done by the ESA showcases humanity's continuing quest and ingenuity to journey into space and form habitats there.

42. SPACEBOT FROM SWITZERLAND

Source: ETH Zurich

Dynamic Walking & Flight Phases

The Spacebot, a quadruped robot, was created by students at the ETH Zurich and ZHAW Zurich, Switzerland. It's designed to travel in low gravity environments like the Moon, asteroids and Mars. It travels by "Dynamic Walking" similar to the travel methods used by the original lunar explorers, the Apollo 11 crew, who used a hopping gait in the low gravity environment of the Moon, 50 years ago.

Dynamic Walker

The bot travels by leaping. All four legs leave the ground and leap to heights over six feet. This allows for full flight phases with all legs off the ground. The lower the gravity, the longer the flying time. The Spacebot is being tested at the European Space Agency (ESA). The Swiss team says the robot basically behaves like a mini-spacecraft. They add, this type of dynamic

walking is more efficient and saves energy in low gravity environments like Mars and the Moon.

Dynamic Bot

The robot has spring-loaded legs that store energy and releases
it for the next leap. It also has a reaction wheel similar to that
used in satellites. The wheel enables the robot to take off and
land safely and get acclimated on the ground for exploring. The
next step is to take Spacebot out of the ESA test labs in the Netherlands where lunar gravity is simulated and into the real world
to take on challenging terrain.

43. ROBOT TO INSTALL TELESCOPES ON THE MOON

Source: Stock image of Moon

Telescopes to Look Deep Into the Universe
A NASA funded lab at the University of Colorado is developing robots to deploy small, highly advanced telescopes on the far side of the Moon. In the next ten years, the NASA team will send a rover aboard a lunar lander spacecraft and place it on the dark, far side of the Moon.

Humans & Machines Working Together
The plan is to set-up a network of small telescopes. The deployment will be performed by the rover's robotic arm. The arm will be controlled by astronauts in an orbiting lunar station called Gateway. Gateway will serve as transit to and from the Moon and as a refueling station for deep space missions.

New Telescopic Views of Deep Space
On the far side of the Moon, the telescope will be free of light

and noise. That will enable a unique and pristine gaze into the deep reaches of space. It's a leading edge example of projects underway by NASA, private companies like Space X and other nations. These missions will open up and change the landscape of the Moon forever.

44. DISTANT DISCOVERY IN SPACE

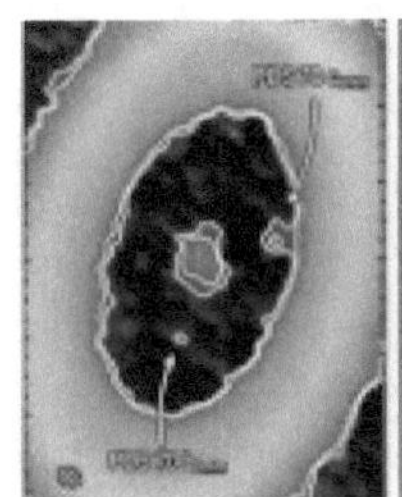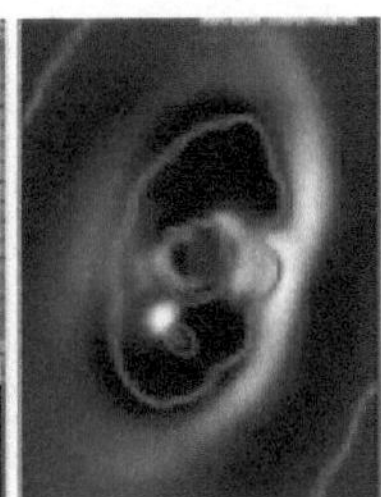

Source: Rice University

World First

Astronomers from Rice University have spotted a moon-forming disk around a very distant exoplanet. The disk, as seen in the color enhanced millimeter wave radio signal on the left, is made of gas and dirt. Scientists say it is similar to the one that is believed to have created the moons around Jupiter four billion years ago. This is a first time observation of a moon forming disk.

World's Most Powerful Telescopes

The Rice University team made the discovery by using the most powerful array of telescopes on Earth. They are located at the ALMA Observatory in Chile.

Exoplanet PDS 70c

The exoplanet is named PDS 70 c. It is a still forming gas giant that's located 370 light years from the Earth.

45. NASA PLANS NEW SPACE STATION

Source: NASA Lunar Orbiter Concept

Man on the Moon in 2024

NASA's plan is for a mini-space station that would orbit the moon with astronauts onboard for 30 to 60 day stays as they travel back and forth to the surface of the Moon. The new space station will be a lunar orbiter, lander and astronaut habitat. It's called the Lunar Orbital Platform - Gateway that NASA wants to use to speed travel and exploration of the Moon. The Trump Administration wants humans on the Moon again by 2024.

Gateway on a Fast Track

NASA wants the 1st phase of Gateway - the Power & Propulsion phase - to launch in 2022. The 2nd phase is the lunar habitat where the astronauts will live, work and do scientific experiments, on and near the Moon. In 2024, it's expected that both components will be launched on a commercial rocket.

International Partners

Canada has already signed on as a partner on the new space station with NASA. Canada is going to build a robotic arm called Canadarm 3. The European Space Agency is also looking to partner. The space station will be equipped to enable the astronauts to do spacewalks. And, NASA envisions that at a later date it will have commercial uses and also be a stopover for astronauts on their way to Mars.

46. SPACE CYBERSECURITY AT RISK

Source: European Space Agency

Royal Institute of International Affairs Report

UK based Chatham House or The Royal Institute of International Affairs has released a report warning that there is an urgent need to address the cybersecurity of satellites in space. They are asking NATO and NATO member nations to address the cybersecurity of space-based satellite control systems because they are vulnerable to cyberattack, particularly if a network system were breached.

International Security

There are important ramifications from this cyber vulnerability finding because virtually all military operations rely on space-based data and communications for instant decision making. According to the research report, any threat to a satellite's control system or available bandwidth "poses a direct challenge to national critical assets" and international security.

Critical Space Assets

With the military dependent on space based architecture, Chatham House researchers say there are new and growing cyber threats that can put military missions in jeopardy. The group is calling for a major investment to harden and protect satellites from hacking so that they can provide reliable and accurate information particularly for the military.

47. NASA SPOTS ELECTRIC SOCCER BALLS IN SPACE

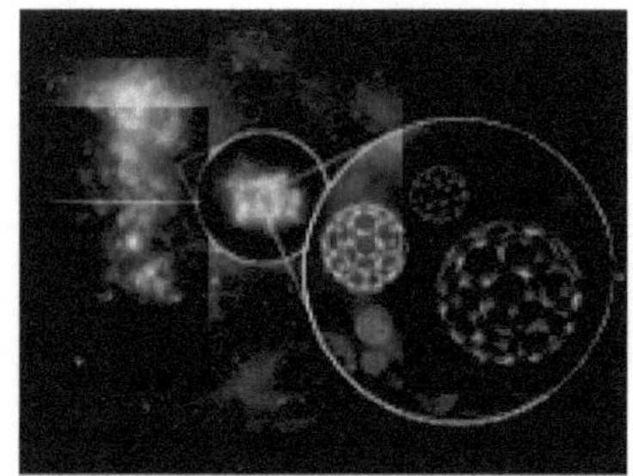

Interstellar Presence

The Hubble Space Telescope has discovered electrically charged molecules shaped like tiny soccer balls in space. They say this is highly significant because it sheds light on the mysterious contents of gas and dust that fill interstellar space, the so-called interstellar medium (ISM). This is the first time that an electrically charged, ionized "buckyball" has been found in the interstellar system.

Buckyballs

The buckyball molecules are composed of 60 carbon atoms that configure like a soccer ball. The molecules are a form of carbon called buckminsterfullerene or Buckyballs after the inventor of the geodesic dome, the inventor Buckminster Fuller. NASA scientists say this discovery sheds light on ISM and the beginning of planets.

ISM

NASA scientists believe that interstellar gas and dust are the starting point of the chemical process that creates stars and planets. Life as we know it is based on carbon bearing material. Now that NASA scientists have discovered that carbon buckyballs have formed and survived in the harsh environment of interstellar space, they want to determine how widespread this might be in the universe.

48. OUT OF THIS WORLD EXOPLANET

Source: NASA Artist Rendering

Exoplanet LHS 3844b

A NASA space telescope has revealed for the first time a rocky Earth-sized exoplanet. It's outside of our solar system and tightly orbits around the most common type of star in our home galaxy, the Milky Way. The exoplanet has no atmosphere to sustain life.

Atmospheric Void

Researchers published their discovery in the journal Nature. They say the surface of the exoplanet, known as LHS 3844b, is most likely barren like the Moon or covered in dark volcanic rock. The exoplanet is 49 light years away from the Earth. It's one of 4000 exoplanets, discovered in the past 20 years, that orbit stars in the Milky Way. It's 1.3 times larger than the planet Earth.

49. UAE'S HOPE MARS MISSION

United Arab Emirates Going to Mars

The United Arab Emirates has big space ambitions and plans. It is going to launch a mission to Mars in the late summer of 2020, making it the first Arab nation to launch an interplanetary mission. The Hope Mars probe will deploy the Hope Satellite to study the climate on Mars.

Destination Mars 2021

When the Hope satellite reaches Mars in 2021, it will show how Mars' climate conditions change throughout the year. It will provide a "holistic" mapping of Mars and orbit the Red Plane for two years. The mission could be extended through 2025. Interesting, a female scientist is a key player behind the mission. Sarah Al Amiri is deputy project manager of the Mars Mission and Chairperson of the UAE Council of Scientists. 90% of the workers on the Emirates Mars Mission are under 35 years old.

Highly Advanced Technology

The Mars probe will carry three highly advanced pieces of equipment. There's the Emirates Exploration Imager, which is a camera that will send high-resolution images back to Earth. The Emirates Mars Infrared Spectrometer will study ice, water vapor, dust and temperature patterns in the Martian atmosphere. And, the Emirates Mars Ultraviolet Spectrometer will study Mars' lower and upper atmosphere and be used to determine what causes hydrogen and oxygen on Mars to escape into space.

Crowded Martian Skies

The UAE Mars team is working with space experts at the University of California Berkeley, the University of Colorado Boulder and Arizona State University. Mars in the 2020's will be a crowded place with four major space missions underway. NASA's 2020 Rover mission, the European Space Agency's Exo-Mars rover mission, China's Mars Explorer and now the Emirates Mars Mission will all be probing Mars, which is 140 million miles from the Earth.

SPACE 2020'S: WHAT'S UP THERE

INTRODUCTION

Have you ever wished that you could journey to a whole new world? My book "Space 2020's: What's Up There?" takes you on journeys to many incredible new worlds in space. The worlds are mysterious, magnificent, beautiful and just being discovered in deep space. In many cases, these celestial worlds being found by powerful space telescopes like Hubble take you to the very edges of distant space never seen before. These are journeys of discoveries in both space and time.

The latest space discoveries that we showcase include baby stars being born, Earth-like planets that could contain some form of life, galaxies light years away with billions of stars, an exotic planet raining diamonds, beautiful and massive supernovas, wrong-way meteorites, space tsunamis, distant planets with double sunsets, black holes, black matter, a star that's older than the universe and so much more. We also chronicle the search for life – extraterrestrial life – in the universe that NASA is spearheading.

EDWARD KANE

These space discoveries and space missions by NASA, the European Space Agency, astronomers at universities and global space agencies, including China, offer time-travel trips to the creation of stars and planets billions of years ago. In some case, these discoveries will carry you to the very dawn of the Universe. These exciting discoveries are re-defining our perceptions of space and pushing deeper and deeper into the farthest reaches of the vast Universe around us.

Get ready for a journey into space!

TABLE OF CONTENTS

AUTHOR'S BIOGRAPHY

Ed Kane created and serves as Executive Producer of CEO Global Foresight, a national news/public affairs program on PBS focused on breakthrough innovations. Ed is a licensed stock broker who worked for Smith Barney. He is the author of 23 books on the latest innovations. Ed is a science graduate of the University of Pennsylvania.

1.NASA Discovers Earth-Like Planet

Source: NASA Artist Rendering

Is There Life Beyond Earth?
Have astronomers found another Earth? A team of international astronomers have discovered a planet the same size as the Earth. Like the Earth, the planet orbits its Sun Star. This exoplanet is called Kepler-1649c.

Habitable
The planet seems to have habitable regions. It is orbiting in its star's habitable zone. That is a region around the star where a rocky planet could support water.

Remarkable Discovery
The scientists discovered the Earth-like planet while using re-analyzed data from NASA's Kepler space telescope. Kepler was

retired in 2018. Previous scans of the data with a computer algorithm misidentified the space object. Astronomers taking a hard look recognized it to be a planet.

Space Specifics

The planet is 300 light years away from the Earth. Of all the exoplanets discovered by Kepler, this one is most similar to Earth in size and in temperature. The amount of light it receives from its star is 75% of what the Earth
receives from the Sun. However one big difference is that the planet's star is a "Red Dwarf",which are known for stellar flare-ups. That could make life on the planet a challenge.

Amazing Discoveries

According to NASA Asso. Director Thomas Zurbuchen,
this discovery gives us greater hope than ever "that a second Earth lies among the stars." He adds, even more amazing discoveries are on the horizon.

2.Tumbling Pieces of Wrecked Planet

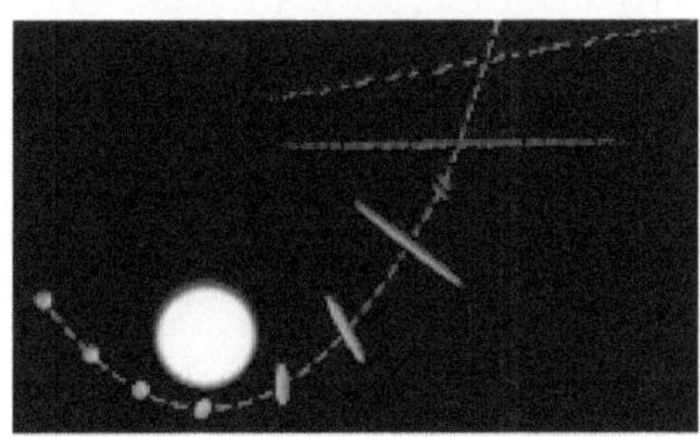

Source: NASA/ESA

Oumuamua Hurtling Through Space

Astronomers call the tumbling space object Oumuamua. In Hawaiian, that means a message from afar. That's what it certainly is. Oumuamua was first discovered in 2017. Some scientists thought it could be a vestige of some sort of alien life, because everything about this space object is so unusual.

New Discoveries and Findings

Two astrophysicists, Yun Zhang at the Observatoire de la Cote d'Azur in France and Doug Lee at U-CAL in the US, have different findings. They believe that the reddish-colored, cigar shaped object tumbling through our solar system is the fragment of a wrecked planet that got too close to the star it was orbiting around. The gravitational force of the star destroyed the planet and the fragment is among the pieces that are left.

First Arrival

Oumuamua is the first object from another star system found passing through our solar system. Fortunately, it's passing out of our solar system and will reach Uranus in August 2020. The fragment is 1/4 miles long. And, it's elongated like a cigar in shape. It has very odd motion and a dry appearance. The scientists say it isn't like an ordinary comet or asteroid.

Computer Simulations

Their computer simulations tell the team that this is the remnants of a wrecked planet, that blew up from the stronger gravitational forces of another object in space such as the "tidal forces" of a star. This is one of the extraordinary new discoveries made in the dynamic universe all around us.

3. Massive Star Explosion

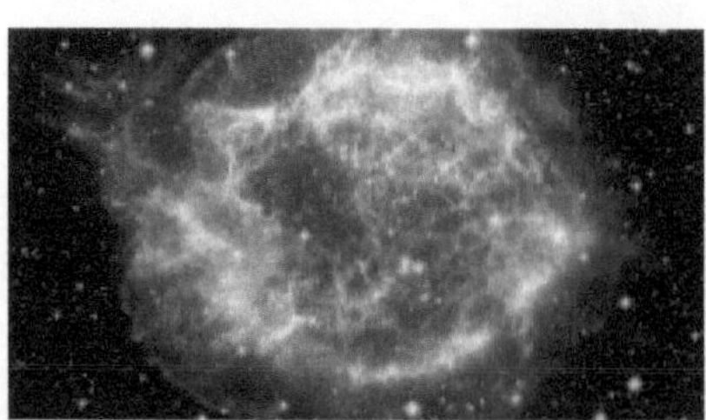

Source: NASA

Most Powerful Supernova Ever

An international team of scientists has discovered the most powerful and brightest supernova ever
recorded. The space event is off the charts. A supernova is the explosive death of a star. In this case, scientists theorize it may have been two stars that merged and then exploded. The supernova event is unique, extreme and one for the record books.

Billions of Light Years Away

The event occurred 3.6 billion light years from Earth. It's named SN2016aps. According to lead astronomer Matt Nicholl of the University of Birmingham, the team measured the supernova in two ways: total energy of the explosion and the amount of light or radiation emitted. With normal supernovas, the radiation amounts to 1% of the total energy. In this one, the radiation is an incredible 5 times the explosion energy.

Bigger than the Sun

This supernova produced the most light astronomers have ever seen. They believe it contained 50 to 100 times the
mass of the Sun. Their report on the discovery has been published in the journal Nature Discovery. The astronomers say they believe even more exciting discoveries
are on the horizon.

4. Hubble Turns 30

Source: NASA, ESA Hubble

New Discoveries for Hubble
The Hubble Telescope has been exploring and discovering New
worlds in space for thirty years. It was deployed in space on
April 25, 1990. More than anything else, it has shown mankind
the Universe around us.

Early Years
The first few years of its space tour of duty were
riddled with technical problems, particularly with
its mirrors. But new technology and new cameras
were launched and deployed by astronauts on Hubble.

Hubble Field
By 1994, it began its riskiest and richest mission - the
Hubble Field. For ten days scientists used Hubble to
examine what seemed like a dark spot in the Universe.
No signs of any brightness, probably no stars. What Hubble
found were thousands of distant stars and
galaxies. Those findings, back in 1994, have forever changed
our view of the Universe, as an immense world of discovery.

What's Next for Hubble
Over the years, Hubble's journeys have moved deeper
into space for wide field, celestial surveys like Hubble's

Ultra Deep and eXtreme Deep Fields. They've revealed very distant, deep in space massive galaxy clusters.

With new instruments, Hubble future missions include WFIRST and LUVOIR which will take the spacecraft deeper and deeper into space. Hubble's legacy and its future is to bring the Universe into closer focus for all of us to discover.

5. Wrongway Meteorite

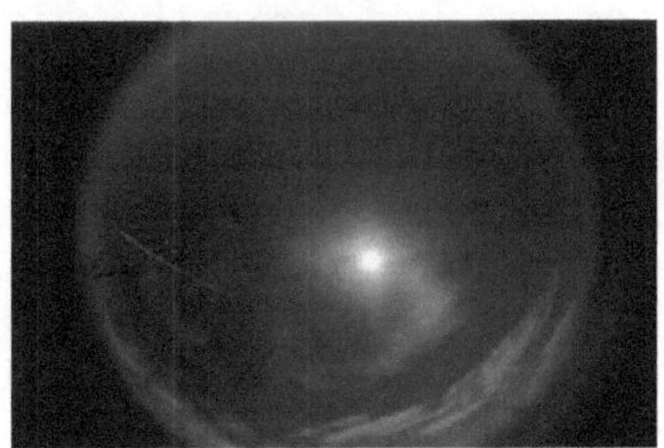

Source: Australia's Desert Fireball Network

Streak of Light the Width of Texas

The celestial event, called a Grazing Fireball, happened over Western Australia. The meteorite that blazed over the night skies took a highly unusual path. For a minute and a half it burned, carving an arc of blazing light as wide as the state of Texas, then it faded away and took back off into space. It's being called the meteorite that ignored the one way signs of our Solar system.

On to Jupiter

What's incredible is this meteorite didn't disintegrate in our atmosphere or crash to Earth. After the light display faded, It took off back into space. According to Australian astronomers and scientists, who studied the highly unusual event which happened in July 2017, the wrong way meteorite is headed to Jupiter. They say, the likely arrival date is 2025.

Blazing Speed

This is an extremely unusual celestial event. And, it's an extremely rare meteorite that grazes into the Earth's atmosphere at a very low angle. It skipped across the skies like a skipping stone on a pond. The Australian team estimates that the meteorite weighs 130 pounds, is a foot across and moves at a blazing speed of 10 miles per second. They believe it originated in the asteroid belt between Jupiter and Mars.

Ignoring the One Way Signs of Space

Grazing fireball events are extremely rare. In 1783, historical records tell us that the Great Meteor streaked across the skies of England and Europe and streaked back to space. There was a similar meteorite, the "Great Comet", that blazed across the skies of the northeast US in 1860. And one was spotted in 1972.

Interstellar Space

When the Australian meteorite reaches Jupiter in 2025, it will have to grapple with Jupiter's gravity. Then it is likely to be tossed into interstellar space. Astronomers are not speculating on the next leg of its quixotic journey. This is the first grazing fireball that's been studied by scientists. And, it's also the one that spent the longest amount of time in the Earth's atmosphere, as far as scientists know.

6. NASA Clocks Massive Winds in Space

Source: NASA

Winds on a "Failed Star" of 1,450 MPH

NASA has achieved yet another first in space. NASA scientists have clocked winds of 1,450 mph on a huge "brown dwarf" or failed star, that is 33.2 light years away and outside of our solar system. The object is the size of Jupiter, the largest planet in the solar system, and 40 times Jupiter's mass. The winds are extremely fierce. This discovery is a first for astronomers. Previously they were only able to measure winds speeds on planets and objects within our solar system.

Whole New Worlds

The scientists say this discovery opens the way to understand atmospheres unlike anything in our solar system. The technique they used to make this discovery is also a first. The discovery was made by combining the detection of radio and infrared emissions from the brown dwarf. They measured the spinning rotation rate of the dwarf and the rotation rate of the atmosphere around the dwarf. By comparing those rates, the NASA scientists made history and establish the wind speed around a huge object in distant space.

7. NASA's Asteroid Defense

Source: NASA

NASA Spacecraft to Collide with an Asteroid
In Spring 2020, NASA began its rigorous process of prioritizing ten years of space targets and tasks for 2023 through 2034. A fascinating new component is the inclusion of planetary defense as a priority. A top example: a planetary defense system against huge asteroid strikes hitting Earth.

DART Spacecraft Defense System
In July 2021, NASA will launch its first planetary defense spacecraft mission called DART, the Double Asteroid Redirection Test. The key word in this asteroid mission is "Redirection" - redirecting the asteroid from a major and harmful hit on Earth. The NASA spacecraft will be programmed to collide with the smaller half of a double (binary) asteroid called Didymos in September 2022. Incredibly the spacecraft will hit the asteroid at a speed of 13,500 mph.

911 Asteroid Patrol??
The purpose of this space venture is to see how the collision

impacts the asteroid's trajectory and path. The ultimate goal is to find a means to protect the Earth from a large asteroid hit by throwing the asteroid off-course into a safe landing spot and also cutting its size. This is not a 911 asteroid response. Experts say they need substantial lead times of 5, 10, 15 years to do this type of tactical space surgery. They need that time to nudge the asteroid and send it on its way elsewhere so it doesn't cause major damage on Earth. The asteroid collision by the NASA spacecraft in 2022 will be tracked and monitored by other satellites including from the European Space Agency.

8. Safer Satellite Fuel

Satellites with Hybrid Fuel - Greener Flights

A global research process is underway to discover new, clean fuel propellant systems to propel the escalating number of discovery, communications and other business satellites being launched into space. The current standard propellant for space satellites is a highly toxic substance based on hydrazine fuel. If this fuel leaks onto the ground on Earth, it is extremely harmful to humans, highly toxic and very explosive. Just consider the exposure to a worker on a NASA or European Space Agency satellite project. They use hazmat suits on the job.

New Sat Propulsion Systems

Global research teams are working on much better and cleaner satellite propellant systems. Discoveries of better solution have been made. But the implementation process for new change is difficult because it disrupts current satellite systems.

Need for Cleaner SATs

The number of satellites being launched is soaring and so are the fuel used in the launches and the propulsion used in satellite orbits. Efforts are underway to develop much cleaner fuel. One company NanoAvionics has developed a less harmful propellant that's based on nitrogen, hydrogen and oxygen. The substance is called ammonium dinitramide and it is very promising in propelling smaller satellites like CubeStats. A pilot project is underway and deployed in space orbit.

Airbus

Airbus has developed a hybrid propulsion system that uses hydrogen peroxide. The program is called HyproGEO and is for satellites in geostationary orbit around the Earth. These GEOsats are big communications and broadcasting satellites that orbit the Earth in a fixed point. The system passes hydrogen peroxide over a catalyser to produce hot oxygen and water vapor. The oxygen provides the propulsion or it can spark another substance for an extra boost. The Norwegian Defense Company NAMMO successfully tested the HyproGEO hybrid engine to launch a rocket. These clean fuel systems are under development. They provide big cost savings. In fact, NanoAviaontics claims its system operates at 1/3rd the cost of current CubSat launches.

9.Black Hole Catching a Star

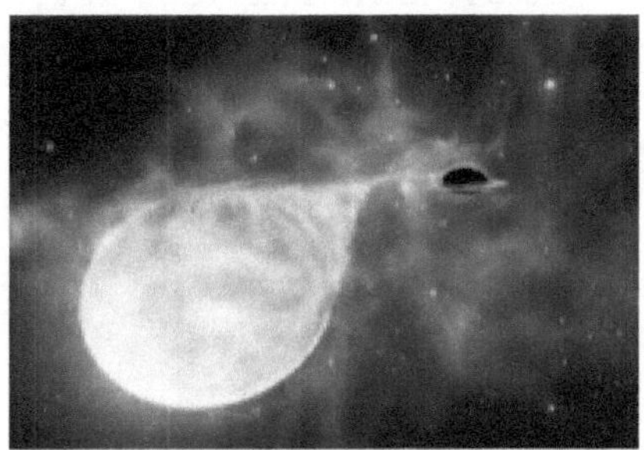
Source: ESA/.Hubble Artist Rendering

Elusive Missing Link of Intermediate Massive Black Holes

The Hubble Space Telescope has detected what looks like a very elusive intermediate-mass black hole in space (IMBH). The Hubble picture captures the black hole in the process of pulling in a passing star. Incredibly, the black hole is 50,000 time the mass of the Sun. If this discovery proves to be true, it will be a breakthrough. It will provide astrophysicists a big jump forward toward the missing link they've been looking for to understand the evolution of black holes. They don't understand how intermediate sized black holes
assemble and if they are the building blocks for massive black holes. This is another example of the dynamic universe that surrounds us.

Tidal Disruption of a Star

Hubble, in its remarkable space journey, caught this image in April 2020. Astrophysicists at the University of New Hampshire in Durham just published their analysis. They describe the Hubble image as the "tidal disruption of a star" by an intermediate-mass black hole. The action took place in a massive star cluster at the outskirts of a huge galaxy. The galaxy is way beyond our Milky Way. According to NASA, this happened deep in space in a distant, dense star cluster.

Why Is This Important?

Why is this discovery so important? Because it has been so elusive and so critical to understanding the evolution of black holes. Intermediate-mass black holes are smaller, less active and much more difficult to detect then massive ones in space. And astrophysicists believe they are a key link in the chain of black hole creations in the Universe.

10. 1400 Year Old Space Mystery Solved

Source: Space

Red Fan Skies Over Japan

More than 1400 years ago, a dazzling red fan spread out across the skies over Japan, according to historical records. The phenomenon has been documented by modern day astronomers and has baffled them ever since. For the inhabitants of Japan in the year 620, the red skies were frightening. They called it a "strange red sign" in the sky. Astronomers and space weather experts, working at the National Institute of Polar Research in Japan, have determined that the red skies were an ethereal sky phenomenon.

Pheasant Tail

The scarlet fan-like light over Japan happened on December 30, 620. Those in Japan who saw it said it looked like a pheasant tail and many considered it a bad omen. For years, modern astronomers have considered it to have been a comet or an aurora. But, neither seemed to fit until now.

Space Weather

Japanese astronomers have solved the mystery. The explanation is space weather and the 7th century heavenly display is a spectacular aurora. This is a space event in which elements of the Earth's atmosphere are being activated by charged particles spit out by the Sun. Astronomers have also discovered more recent displays of auroras over Japan that are pheasant tail shaped and scarlet red like the aurora in 620. The 1400 year old space mystery has now been solved.

11. NASA's Solar Mission of Discovery

NASA's Image of Solar Particle Storm Spewing Plasma into Space

Six Tiny Satellites Form One Massive Telescope

NASA's is using unprecedented technology for discoveries in its next journey to the Sun. NASA has plans to deploy the largest radio telescope to ever fly into space. The system contains six tiny satellites that work in sync, fly in tanden and come together like a tech rock band to form the world's biggest telescope in space. The goal is to investigate powerful Solar storms. The mission is called SunRISE, or Sun Radio Interferometer

Space Experiment. It is designed to help scientists significantly upgrade information on the Sun's dangerous activity around the Earth called Space Weather. NASA hopes to launch the satellite system in 2023.

Space Weather

Understanding and predicting space weather events are very important because of future plans for ordinary citizens to start travelling into space to the Moon, Mars and beyond. The new satellite radio telescope system will start in a low Earth orbit. NASA's goals are to understand the Sun and how it erupts with space weather events. The hope is that experts will learn how to mitigate the potentially dangerous effects of space weather on space travel, space crafts, astronauts and travelers in space.

12. Laser Guide Star

Source: European Southern Observatory

Laser Beams Over the Milky Way

The technologies behind new discoveries in space are amazing. An example: the European Southern Observatory's Very Large Telescope (VLT). In the night skies over Chile, the VLT beams its laser guide star. The beam of laser light from the laser guide star stretches over the Milky Way Galaxy. Astronomers are using giant laser beams to correct their highly powerful telescopes from distortions caused by turbulence in the Earth's atmosphere. The turbulence causes stars to appear to twinkle when we on Earth star-gaze at night.

Space Discoveries from Chile

The laser guidance system for powerful, advanced space telescopes discovering new exoplanets, galaxies, stars and more is based at the Laser Guide Star Facility at the ESO's Paranal Observatory in Chile. Some of the most powerful telescope systems in the world for peering into space are located at the Observatory.

Search for Habitable Planets

Paranal is also home to a huge, 4 telescope system called SPECULOOS or Search for Habitable Planets Eclipsing Ultra-Cool Stars. The system watches for dips in the brightness around "ultra-cool" stars, similar to brightness dips when the Moon eclipses the Sun. The image below is the view SPECULOOS has provided of the Carina Nebula in deep space. The Carina Nebula is a nursery for baby stars.

Source: ESO image of Carina Nubula

13. $10 Billion Space Exploring Telescope

Source: NASA's Webb image

Hubble's Incredible Replacement Partner in Space

The $10 billion James Webb Space Telescope "Webb" will launch in March 2021. Its mission of space discovery is awesome and its technical capabilities are intergalactic. Webb will give all of us on Earth a brand new look at the massive, dynamic mysteries of the cosmos around us.

Hubble Successor to Discovering Space

This is the highly advanced successor to the amazing Hubble telescope. Webb has advanced technology to detect infrared light that enables it to look further back in time than any other telescope. It will explore the mysteries surrounding the creation of the Universe, photograph and image distant galaxies, stars and exoplanets and further understanding of the dynamics of the solar system.

Webb is Unique

Webb is the most advanced, complex and ambitious space telescope ever developed. It's a joint venture between NASA, the European Space Agency (ESA), the Canadian Space Agency (CSA) and the Space Telescope Science Institute. The telescope is massive and contains an infrared camera and spectrograph to collect light from the early beginnings of the Universe. It is able to see much deeper back into time than any other space telescope. It has the capability of imaging the dawn of the cosmos, detect first ever images of stars, galaxies and exoplanets. Its instruments can block the light of a star in order to detect distant exoplanets and their atmospheres. This will be a time tunnel journey into the beginnings of the Universe that is soon to unfold.

Building On Hubble History

Hubble has made enormous space discoveries in the past 30 years and into 2020. By comparison, it is a $2.5 billion, 11 ton telescope launched by NASA in 1990 with a primary viewing mirror diameter of 8 feet by 2.4 meters. Webb is a 6 ton telescope with a primary viewing mirror diameter of 21 feet by 6.5 meters with greatly advanced infrared resolution. The new views from Webb will redefine the universe as we know it.

14. Planets With Double Sunsets

Source: NASA Artist image of 3 planets & 2 stars in the Kepler-47 system.

300 Exoplanets With More than One Sun-Star

New research shows that double sunsets on planets may be just as common as the one sunset a day we enjoy on the planet Earth. Astronomers say that has important implications for the search for other forms of life outside of our Solar System. New research has discovered more than 300 exoplanetary systems that have more than one Sun-Star and double or more sunsets per day. This is Star Wars come alive, with Luke Skywalker's Home planet Tatooine showcasing its double sunsets in iconic images in the George Lucas movie series. Only now there's proof Tatooine type planets exist in the Universe.

Where There Are Twin Suns, Could There be Life?
Astronomers prioritize Sun-like stars as the most probable places in the universe where alien life could exist. For massive stars that are most like the Sun, new research tells us that 50% of them exist with companion stars. Astronomers are focusing on these binary stars to see if they are hosting Earth-like planets. That possibility has almost been ignored until now. The change of focus happened when a team of astronomers using data and observations from Europe's Gaia spacecraft discovered more than 300 distant planets with binary Sun-stars. The game-changer is the exponential innovation in new technology to explore distant space. The search for life in these exoplanets is starting to unfold.

15. Pink Cloud of Baby Stars

Source: NASA's Hubble

A Star is Born
The Hubble Space Telescope has discovered and provided spectacular images of a dazzling pink cloud that is a birthplace in the universe for massive stars. The cloud is known as LHA 120-n

150. It's a cloud of dust and gas surrounded by sparkling young stars. You might call it a "space nursery" for baby stars.

Distant Views into Space.

The pink cloud is located near the Tarantula Nebula, which is a known hot spot for star formation. The images that Hubble is providing to astronomers will help them learn how massive stars are created. Astronomers' theoretical models indicate that massive stars should form within clusters. But astronomers are learning through observation that at least 10% of massive stars form in isolation in space.

Time Tunnel into the Past

Hubble is a joint project between NASA and the European Space Agency. It has provided humanity with the most distant views into space ever achieved, including images of 5,500 new galaxies. Some of the galaxies are nearly as old as the universe, which is estimated to be 13.7 billion years old. The galaxies are 13.2 billion light years away, meaning their light takes 13.2 billion years to travel to Earth. NASA calls the images that Hubble has and is providing "a time tunnel into the distant past." The images take us back more than 13 billion years in time.

16. NASA's Uranus Discovery

NASA Spacecraft Flew Right through It

NASA researchers have made a remarkable discovery from the Voyager 2 space mission to the outer planets more than thirty years ago in January 1986. They've spotted a massive bubble breaking away from Uranus' atmosphere that may have taken some of the planet's gassy atmosphere with it. NASA announced the discovery in 2020. They're calling it a huge plasmoid loaded with supercharged particles. Remarkably, the Voyager 2 spacecraft flew right through it.

Plasmoids

Plasmoids are massive space bubbles filled with charged particles that can break away from a planet. They pull charged particles out of a planet's atmosphere and hurl them into space. Such activity has been spotted from Earth and nearby planets but never before on Uranus. NASA scientists are studying this event because, they say, if you change a planet's atmosphere, you can change the planet itself.

Huge Space Event

NASA believes this plasmoid was huge. They estimate it as 250,000 miles wide and 127,000 miles long. The NASA scientists discovered it by analyzing the magnetic field around Uranus as captured by Voyager 2, every two minutes. No research had drilled down on the magnetic field to that degree. Their analysis revealed an abrupt zigzag in the magnetic field that lasted one minute during the Voyager's 45 hour journey past the planet. That one minute wobble in the planet's magnetic field was a huge plasmoid breaking away from Uranus like a massive teardrop and hurling energy into space NASA believes that learning from such events may help scientists to better understand the formation of planets.

17. Robotic Satellite Repairman

Source: NASA

New DARPA Space Initiative

The US Department of Defense Advanced Research Projects Agency, DARPA, has launched an innovative program to extend the life of US satellites orbiting in space. In essence, DARPA is creating a robotic system to inspect, service, repair, improve and extend the lives of US satellites in orbit. The program promises significant money savings and much longer and efficient use of satellites.

Space Logistics

DARPA is partnering with Space Logistics, a subsidiary of Northrup Grumman, to develop and deploy advanced robotic capabilities in space. The first step is to develop what DARPA is calling a "DEXTEROUS ROBOTIC SERVICE". The technology needed would increase satellite life spans, improve their reliability and resilience. All of which would save the US government and space companies significant money lost when a satellite's operations breakdown and are currently unfixable in their space orbit.

In GEO

The robotic servicing system would be in GEO or geosynchronous orbit, as are the satellites that it will be servicing. Space

Logistics will build the robotic servicing spacecraft, integrate it with a launch vehicle, do the launch and then offer commercial satellite servicing to government and commercial client spacecraft. This promises to be a much needed maintenance and repair service for satellites.

18. Superconducting Material Discovered in Meteorite

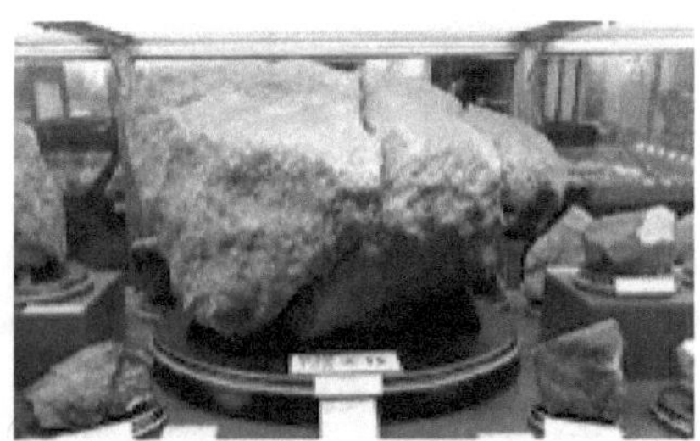

Super: Mundrabilla Meteorite

First Evidence of Superconductivity in Space

There is superconductivity in space naturally occurring. That's the shocking discovery made by scientists at the University of California San Diego. The UC team led by Dr. Ivan Schuller found trace amounts of superconducting material in one of the world's largest meteorites, Mundrabilla that landed in Australia. It weighed 22 metric pounds and broke into pieces on impact. The superconducting material found in Mundrabilla is a known alloy of indium, lead and tin. But the fact that it is contained in a meteorite came as a shock. This is the first evidence of superconductivity in space.

Coveted Material

Superconductors are materials that conduct electrical current without any resistance. They are coveted by researchers developing quantum computers and by companies wanting to transfer energy with much greater efficiency. The UC team has devised a method to quickly scan materials to see if they are a superconductor or not. They are looking for superconductors everywhere, including in materials from space.

19. A Selfie from Mars

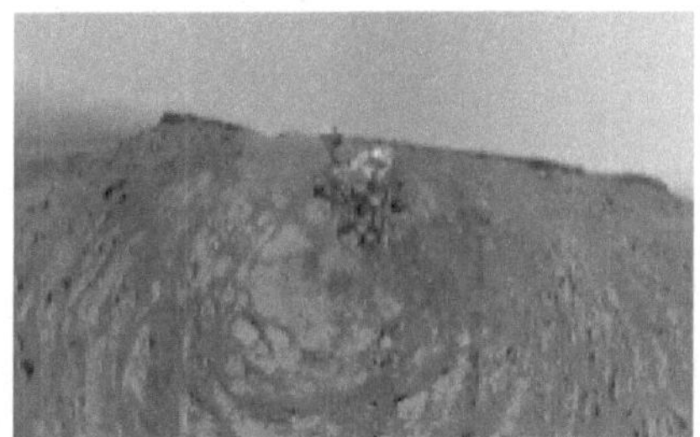

Source: NASA, JPL Caltech

Space Shots

NASA's Curiosity Mars rover snapped a picture of itself after drilling a hole into a rock structure on Mars. The picture documents the tough job in rough terrain that this remarkable piece of robotic technology is performing. Curiosity made its way up the steep rocky incline called the Greenheugh Pediment, as seen in the picture behind the rover.

Space Renaissance 21st Century

This is part of an 86 image series taken by the camera on the Curiosity Rover's robotic arm. Curiosity has spent more than six years roving the harsh and mysterious environment of the Red Planet and sending back images and samples to NASA. Part of that mission is to determine if life ever existed on Mars and if deep within the surface of Mars there is water that could sustain life.

20. NASA Discovery – Massive Space Tsunami

Source: Hubble image of quasar tsunami

Energy Outflows Accelerating at 46 Million Miles Per Hour
NASA's Hubble space telescope has spotted tsunamis ripping across galaxies in deep space. The tsunamis spotted by NASA astronomers are actually quasar tsunamis. Quasars are objects similar to stars in distant space. NASA is calling it "the most energetic outflow ever witnessed in the Universe."

13 Quasar Outflows Spotted
The outflows from the quasars tear across interstellar space causing massive havoc to the galaxies where the quasars reside. The amount of energy is tremendous. And the velocities of the quasars is unfathomable - 46 million miles per hour, or as NASA puts it a few percentages of the speed of light. NASA was able to study 13 quasar outflows and also measure the speed of the gas spewing out of them being accelerated by the quasar wind.

Quasars, Black Holes and Galaxies
Quasars contain massive black holes fueled by falling matter from within them. As the black holes devour the matter, hot gases and radiation are emitted in massive amounts. Winds driven by the radiation pressure push material from the galaxy's center. The outflow tsunamis accelerate to breathtaking speeds. And the light created is incredible. The quasars shine 1000 times brighter than the host galaxies that are comprised

of billions of stars.

21. Basic Element of Crude Oil Discovered on Mars

Source: NASA image of Mars

Where There Is Oil, There Could be Life

Two astrobiologists have found the basic elements of oil on Mars. Jacob Heinz of Berlin's Technische Universität and Dirk Schulze-Makuch of Washington State University in Saint Louis have found organic compounds called "thiophenes" that are present in crude oil and coal on Earth. The thiophenes were discovered in dried mud that NASA's Curiosity Rover dug up in Mars' Gale Crater. This is a remarkable scientific discovery that raises the intriguing question, once again, about life existing on Mars.

Oil and Life

Where there is oil, there many times is life. The issue now confronting the scientists is how the thiophenes on Mars were created. Most often the creation of thiophenes is a "biotic" process that involves living elements. If that is the case, according to scientist Schulze-Makuch, "there was early life on Mars....and there could be extant life on Mars today."

NASA's Curiosity Rover

NASA's Curiosity Rover has been on the ground on Mars for more than six years. It digs up soil, analyzes it and then beams the raw data back to Earth. This new discovery by the two astro-biologists is from dirt they analyzed from Curiosity. Heinz and Schulze-Makuch believe the origin of the thiophenes is biological rather than "abiotic" and chemical from such an event as a meteorite crash. They are working to prove their theory that this is another example of life in space. Also, a team of scientists from Harvard have found proteins on a meteorite, suggesting that life can evolve in the vastness of Space.

22. Autonomous Cars Guided by Satellites

Source: Satellite in space image

Your Car in Contact with Space

Chinese automaker Geely has a space-based plan for "steering" autonomous cars. It is in the process of building its own network of satellites to provide advanced geolocation and connected vehicle functions for self-driving cars. This is a first in China and Geely's satellites will be solely used to support its autonomous driving vehicles. It's a very innovative concept to use satellites to track, communicate with and help guide self-driving cars.

Global Operations

Geely is a huge automotive player. It owns Sweden-based Volvo cars, US flying car startup Terrafugia, Malaysia's Proton and has a $9 billion stake in Daimler, the owner of Mercedes Benz. Geely hopes to share and benefit from Daimler automotive technology. Geely is integrating Volvo with Geely operations, which will create a huge, global automotive brand. Now, Geely is kicking off its satellite network with a $325 million investment to build a development center and factory to manufacture the satellites in Taizhou city.

Low Orbit Communications Satellites

Geely's satellites will be low orbit communications sats for geolocations and to support connected functions for self-driving vehicles. This is reminiscent of Elon Musk's two main businesses, the electric vehicle giant Tesla and his powerful space company, SpaceX. Musk has successfully combined synergies from his automotive and space businesses. It appears Geely has a similar concept in mind. Musk's Tesla operations have been rapidly growing in China.

23. Adidas Marathon in Space

Source: NASA ISS

Putting Running Tech to the Longest Distance Test

Adidas is really going the distance. It sent running shoe material aboard a SpaceX rocket to the International Space Station. The purpose is to develop the most comfortable, high performance running shoes from experimentation in the harsh conditions of space. The German based footwear giant is performing the first ever footwear research in space.

One Big Boost in Zero-Gravity

Boost pellets made of the same plastic that Adidas integrates into the soles of its Boost sneakers and other running shoes are undergoing scientific experimentation in space. This will give Adidas the opportunity to test their performance and maximum configuration in sneakers without the interference of gravity.

World's Longest Scientific Marathon

The plastic pellets are made of two difference polymers with slightly different molecular structures. Astronauts will watch how they perform in microgravity. Adidas says the experimentation in space will give its engineers, scientists and designers a much better understanding of the materials. They also expect to reap sustainability benefits for their circular manufacturing process. For running shoes, this may be the world's longest scientific marathon.

Adidas In Orbit

Adidas is big on doing research in space and its zero-G conditions. It recently tested its soccer ball on the International Space Station. To celebrate its multiyear contract with the ISS Labs, it has launched a line of "Space Race" sneakers. Adidas is "all-in" on the benefits and potentialities of doing research science in space to run faster and better on Earth.

24. Planet Discovered with Exotic Rain

Source: NASA

Exoplanet WASP-76 B

Some scientists are calling WASP-76 B the planet from Hell. It was discovered in another solar system and is therefore an exoplanet. The exoplanet is so exotic that even the weather conditions are extremely bizarre. On WASP-76 B it's usually cloudy with a chance of rain. But the rain drops are hot liquid molten iron that are boiling hot. European scientists and astronomers spotted it using the planet hunting instrument EXPRESSO on a huge international telescope based in Chile. The weather conditions on this exoplanet are so severe there is molten iron rain! This is a unique finding in the Universe. And despite the extremely harsh environment, you have to wonder about the value of the mineral deposits on such a faraway frontier in space.

640 Light Years From Earth

This very hot exoplanet is 640 light years away from the Earth. And, it's a big one. Twice the size of Jupiter which is our solar system's biggest planet. WASP-76 B is one of the most extreme exoplanets ever discovered in terms of climate and chemistry. It's part of a family of exoplanets, discovered in recent years, called "Ultra-Hot Gas Giants".

Scorched By the Sun

The exoplanet receives 4,000 times the solar radiation than we do on Earth. The "sunny side" of the planet heats to 4,350 degrees F which vaporizes any metals present. Strong winds carry the iron vapor to the exoplanet's cooler side where it condenses into liquid iron rain. Scientists from the University of Geneva are leading this exploration. They think WASP-76 B provides a unique living platform to test climate models and to understand extreme atmospheric evolution.

25. Saving Wine From Climate Change

Source: NASA SpaceX Launch 3/6/2020

Vino in Space

A SpaceX launch in March 2020 contained experiments designed to save wine from the destruction of Climate Change. Global warming is putting grapes and wine at risk as they are very susceptible to changes in temperature. The Climate Change situation is so significant some vineyard owners are moving their operations to colder regions in higher altitudes. The wine experiments that SpaceX carried to the International Space Station are designed to remedy the situation.

International Space Station

European startup Space Cargo Unlimited is the force behind the wine experiments. Their project was ferried by SpaceX to the International Space Station. Included are vines for Merlot and for Cabernet Sauvignon to see how they fare during six months aboard the ISS in space. The grape vines will be subjected to zero gravity and increased radiation. Space Cargo Unlimited scientists believe this will result in mutations in the vines that occur quicker than on Earth. It will allow the scientists to track the evolution and determine how the vines make themselves hardier and more resistant to harsh conditions.

Agriculture of the Future

The goal of this effort is to make it possible to produce wine in harsher environments. The vines will be stored in an environment on the ISS similar to wine storage facilities on Earth. The temperatures will be as low as 33 degrees F and the humidity will range from 70% to 80%. Space Cargo Unlimited says they are pioneering a new way to study microgravity and accelerated evolution in plants. They believe it could be a game-changer and unlock the agriculture of the future.

26. Is This the Universe's Dark Matter?

Source: European Space Agency/Hubble

Big Bang Subatomic Particle

Researchers believe that they have identified a subatomic particle that may have formed the Universe's dark matter right after the Big Bang, nearly 14 billion years ago. They say this is one of the strongest pieces of evidence of the very existence of dark matter. Dark matter can be seen in the picture (above) floating like a ring in the galaxy cluster ZwC10024+1652. The team of astronomers, studying images obtained from the Hubble telescope, believe that dark matter was produced by a collision between two huge space clusters.

Dark Matter

Scientists have estimated that 27% of the matter in the Universe could be dark matter. There has been very little substantive understanding of what dark matter is until now. A team of nuclear physicists think that dark matter could be made from the newly identified particle - the d-star hexaquark.

Quarks - Exotic Particles

The smallest level that matter can be broken down into is the quark. Quarks are smaller than subatomic particles, atoms and molecules. The d-star hexaquark is made up of six quarks. Quarks are very different and can combine in very unusual ways. Some researchers are theorizing that hexaquarks could have condensed into dark matter in the unique conditions 14 billion years ago right after the Big Bang occurred.

27. SpaceX's 70 Space Missions Per Year

Source: SpaceX

Starlink Satellite Megaconstellation, Customer & NASA Launches, Starship to Mars

SpaceX and CEO Elon Musk have an ambitious plan for at least 70 space mission launches a year starting in 2023. There will be 38 launches by SpaceX in 2020. So, the plan is to nearly double that. The launches will take place from Cape Canaveral and the Kennedy Space Center in Florida. A big part of the launches will be building SpaceX's Starlink mega-constellation of satellites for global internet services. SpaceX is also enjoying a growing number of space flight bookings from customers and NASA.

Mobile Service Tower For Launches
Space X's Falcon 9 and Falcon Heavy rockets will have more versatility and options with a new mobile services tower that SpaceX is deploying. The tower will allow missions to be assembled vertically rather than horizontally, which is a big advantage. SpaceX will also have the new capability to launch to polar orbit from Florida. And, it's pushing for more space vehicle reusability to cut down on expenses.

On to Mars
Another part of the SpaceX business growth plan is its next generation Starship that the company expects will carry thousands of humans to Mars. In 2020, SpaceX plans to launch and test Starship up to an altitude of 12.4 miles. Starship is designed to be reusable and to go into orbit around the Earth, Moon and Mars.

28. Snowman Star Born by 2 Stars Merging

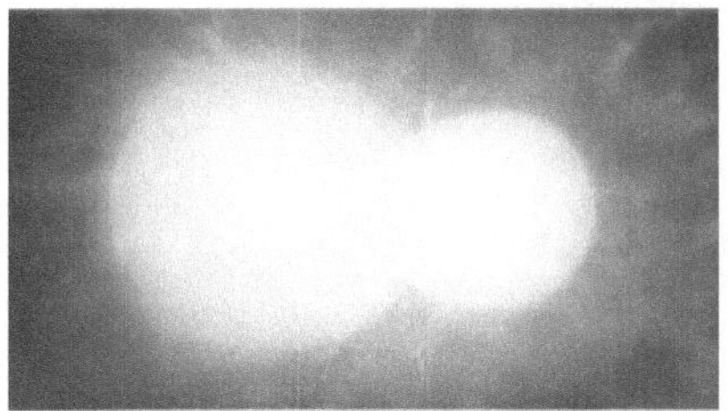

Source: Warwick University, UK

Unique Shape and Unique Atmosphere

Nothing like this new star has ever been discovered in space before. Warwick University scientists have spotted a huge, snowman-shaped star. It's a new star just born by the merger of two "white dwarf" stars. They've named the star WDJ0551+4135. The star's atmosphere is unique and unusually rich in carbon. Everything about this star has the potential of leading to new scientific discoveries about the universe and the birth and death of stars.

Bigger Than the Sun

The new Snowman star is more massive than the Sun. Another unusual aspect of the birth of this star is that the merger of the two "white Dwarf" stars didn't result in an explosion and powerful supernova. The Warwick University scientists say what is particularly exciting for their research is that the stars merged rather than exploded, as is usual.

How a Star is Born

The scientists believe that measuring the properties of the failed supernova will tell them a lot about the "pathways to thermonuclear self-annihilation" of stars. A "white dwarf" is what stars like the Sun become after using up all of their nuclear fuel. In this case, a new star was born by the successful merger of two starts.

29. International Space Station Destination

Source: Axiom

NASA Partners with Axiom

Space exploration company Axiom of Houston, Texas is partnering with NASA for a big launch of space tourism. Tourists will be able to visit the International Space Station (ISS) as early as 2024. Axiom is a private US space station manufacturer that is building, what it hopes will be, the world's first commercial, international space station. CEO Michael Suffredini is the former program manager of the ISS for NASA. This is a big business alliance, with tremendous expertise behind it, for launching tourists into space.

Advanced Luxury Space "Hotel" Priced as High as $35,000

The Axiom space vehicle is an impressive piece of technology. First of all, it will be led by a trained astronaut, have a crew and also research labs for discoveries on board. It has the biggest space observation window ever built into a space capsule and the state-of-the art interior was designed by the top French-based hotel and yacht designer Philippe Starck. This takes space tourism to a whole new level.

Space Nest

The interior is designed like a "nest" according to Starck with hundreds of nanoLEDs that change color and embedded touchscreens and handles. It will also have Wifi. Starck said the space provided him as a designer and the future occupants "multidimensional freedom" as it is zero-gravity with no horizontal, vertical or diagonal limitations.

A Ticket Takes Money & Fitness

The Axium-NASA module is designed for fully trained astronauts whose nations aren't members of the ISS and for private citizens. If you can pay the price, you also need to be very fit to get a seat. Travelers to space must clear a physical and then 15 weeks of very rigorous training that includes jet flights, extreme environment endurance and suborbital space flights. If you clear, you get one ticket to ride.

30. China's Space Economic Zone

Source: Chinese Academy of Sciences - Far Side of Moon with Earth in the Distance

Trillions of Dollars at Stake

In 2020, China secured a leadership role in space exploration. It made the first landing on the far side of the Moon with Chang'e-5Ti, its mission service module. Since that space milestone, China has stated that it wants to establish a "Space Economic Zone" between the Earth and the Moon. From mining minerals to establishing in-space solar power, the zone is projected by Chinese officials to be capable of generating $10 Trillion a year for China by 2050.

Specific Goals

News of China's goals and intent came in a speech given by a top official with China Aerospace Sciences and Technology (CASC), which is China's main, state-run space contractor. To reach its space goals, China intends to develop low cost launch and space vehicles to exploit the economic potential of space. CASC wants breakthrough space travel and exploration technology by 2040.

Space Planes and More
China's space ambitions are big and bold. The goals include development of a "nuclear space shuttle" by 2045, reusable launchers by 2035 and space planes. All to capitalize on China's hoped for "Space Economic Zone". They also plan on capitalizing and exploiting space-based solar power. Clearly, China is looking to profit on space. And, the potential for China and other space leaders is huge ranging from asteroid mining to water and rich mineral deposits.

31. Beyond the Big Bang

Source: NASA

Huge Explosion in Universe From Light Years Ago Detected
Astronomers using data from the most advanced NASA, European Space Agency, India and Australia telescopes have discovered the biggest explosion that the universe has ever experienced since the Big Bang. It happened in a galaxy 390 million

light years away from the Earth. And incredibly it took place in slow motion over hundreds of millions of years. The science to discover this phenomena is remarkable and the details emerging on events going back into space from such a distance and time are amazing, innovative scientific discoveries.

Supermassive Black Hole in the Ophiuchus Galaxy Cluster
The global team of astronomers say the explosion they detected released five times more energy than ever detected before. The explosion came from a supermassive black hole in a galaxy called the Ophiuchus Galaxy Cluster nearly 400 million light years from the Earth.   The power of the explosion was enough to punch a hole into the gas surrounding the black hole that is big enough to contain 15 contiguous Milky Way galaxies. This is an amazing scientific discovery through breakthrough telescope innovation. What it means to the formation of the Universe and Earth, as we know it, is the next chapter of the history of the Universe.

32. NASA Discovers Mars is a Dynamic Planet

Source: NASA

Hundreds of Marsquakes in the Past Year
NASA's robotic lander InSight has delivered some big discovery

and innovation dividends. InSight has detailed, for the first time, that the Red Planet, Mar, experiences quakes like the Earth and the Moon. NASA has called the shakes, rattles and rolls on Mars "Marsquakes". Incredibly, InSight has detected 450 seismic events on Mars in the past year as it has explored Mars on the ground. 20 of the quakes were significant and in the 3 to 4 point magnitude range.

NASA First

According to NASA, this is the first time that they've established that Mars is a seismically active planet. InSight's observations will help scientists better understand how rocky planets like Mars and the Earth form and evolve. And the lander's sensors also detected significant winds swirling around Mars. In fact, there are thousands of passing whirlwinds whipping around Mars.

Mars Quakes

Compared to quakes on the Earth and the Moon, the Mars quakes are relatively small. Mars is more seismically active than the Moon but less so than the Earth. The Marsquakes provide NASA scientists with new data and information on the interior core of Mars. NASA says the quakes on Mars are caused by the long-term cooling of the planet that makes it contract and facture.

Tech Wonder

InSight is a robotic technological wonder. The lander is equipped with heat flow probes to take the planet's temperature, sensors to gauge wind and air pressure, seismometers to detect quakes and a magnetometer to detect magnetism.

33. Is It a Flying Car or a UFO?

Super: Stock Image of UFO

"Alien Vessel" Caught on Tape Over a Car

This is a strange case of a fast flying object - white with a cloud-like appearance - that a driver accidentally taped on his smart phone through the car's open sunroof. He turned over the tape and reported the incident to California authorities. The incident happened in Santa Clara, California on a highway close to San Francisco Bay. The question is what was it: a flying drone, flying car or a UFO?

Caught on Tape

The object moved with great speed and is only visible for a second or two. The object is flat and elongated. UFO expert Scott Waring of ET Data Base examined the tape and concluded that it's very similar to other alleged UFO's sighted over the years around the world. He adds "there's a very real probability that it's a real UFO". He speculated that the "alien vessel" might have come from an underwater base in San Francisco Bay. This is the latest sighting of alleged UFO's from outer space.

34. NASA and ESA's Blackbird Mission

Source: NASA Launch of Solar Orbiter

Unprecedented Solar Orbiter Probe

NASA and the European Space Agency have successfully launched the Solar Orbiter on a journey to the Sun, to take unprecedented views of its blazing solar poles. The Solar Orbiter took off on top of an Atlas 5 rocket. The Orbiter separated from the rocket within 53 minutes to go into space flight as planned and is in communications with NASA. This unique mission is going very well as it moves along on a very long journey to the Sun.

Mission Unprecedented

This US and European mission is unprecedented. It's expected to help scientists understand how the Sun's massive amount of energy affects humans in space and how it impacts all of us on the Earth. The mission will provide unique images, views and data of the sun's blazing pole regions, which are critical sources of data.

Ten Year Journey

The orbiter is equipped with awesome equipment including an array of solar panels and antennas. The journey to the Sum will take ten years. The orbiter will position at 26 million miles from the Sun, which is 95% of the distance between the Earth and the Sun. It will map the Sun's poles which have a concentrated source of solar winds. Those winds are high impact. They penetrate our atmosphere and even can impact satellites.

Main Goal

NASA and ESA scientists want to determine how the Sun creates and controls the heliosphere, which is the massive bubble of protection that surrounds the solar system. They want to know why the bubble changes over time. NASA and ESA scientists believe the answer may be found in the Sun's poles. Those views will first be available from the probe in 2025. Exciting new science for all of us to observe from Earth!

Key Goals to Get This Off the Ground

The key to this mission, that was first proposed in 1999, was to develop a thermal protection system that can withstand the intense heat of the sun. The Orbiter was built in Europe. It's a great example of US-European space exploration. The cost of the Solar Orbiter hurtling into space is $1.5 billion.

35. Stardust Discovery Oldest on Earth

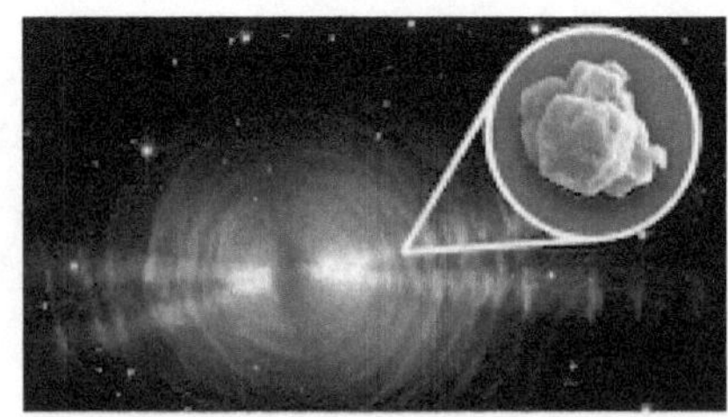

Source: NASA Meteorite

Stardust Material That's Older than the Sun

This is an extraordinary scientific discovery announced by a team of astronomers led by University of Chicago Professor Dr.

Philipp Heck. In a meteorite that fell from space in Australia 50 years ago, there is awesome stardust dating back 5 to 7 billion years. This stardust predates the Sun. Heck and his team analyzed the contents of the meteorite. They found it to be the oldest, solid material ever found on Earth. Their work has been published in the prestigious Proceedings of the National Academy of Sciences.

What This Discovery Opens Innovative Views On
The scientists say this stardust will tell us how parent stars were formed in our galaxy. It will tell us about the origins of the oxygen that we breathe and provide the opportunity to trace materials back before the creation of the Sun. The stardust is the oldest material discovered to reach the Earth. The material is older than our solar system. The Earth is estimated to be 4.5 billion years old. The Sun is 4.6 billion years old. The newly found and analyzed stardust is 5 to 7 billion years old. This discovery does give a whole new perspective on our brilliant and beautiful universe!

36. Saturn's Titan Moon and Life

Source: NASA Jet Propulsion Lab

Moonscapes, Science and Life
NASA scientists have mapped Saturn's exotic, largest Moon

Titan. Apparently, Titan has quite a story to tell about the possibilities of life. NASA has unveiled the first global geological map of Titan. It shows a strange and exotic world that NASA scientists say is a strong candidate for the search for life beyond Earth.

Incredibly Interesting Terrain

The map shows dunes of frozen organic material, vast stretches of plains, and lakes and seas filled with liquid methane...all of which may harbor forms of life beyond Earth. The map is based on radar, infrared and data from NASA's Cassini spacecraft that studied Saturn and its moons from 2004 to 2017. The data and images now emerging are fascinating.

What's Next?

Clearly, Titan contains organic material that is the ingredient critical to form life. What's next is NASA is going to launch its Dragonfly Mission sometime in the 2020's to reach Titan by 2034 with a multi-rotor drone to explore this exciting new Moon in space.

37. China's 2020 Moon Mission on Target

Source: Chinese Mars Lander Test

China - Emerging Space Power

China has successfully completed a landing test of its Mars

lander that will contain a rover to explore Mars. This puts China squarely on track for its Mission to Mars late in 2020. This mission involves landing on and exploring the Red Planet. Mars is cold and barren but it's considered the only planet other than Earth that is habitable. The Chinese space mission could contribute to the international effort to learn about and start understanding the climate conditions and mineral resources on Mars.

Billions of Dollars Invested in Space

China is pouring billions of dollars into its military-run space program. It hopes to have a crewed space station in orbit by 2022. China's space investment is paying off. Early in 2019 it landed a rover on the far side of the Moon, which is a first. Clearly, China is emerging as a major space power along with the US, Russia and Europe.

Safe Landing

In the test, China's refrigerator sized lander was lowered from a crane on 36 cables into Mars like gravity conditions. Using its onboard jets, the lander found a safe place to land amid piles of rocks. The test demonstrated that the lander can successfully maneuver to avoid obstacles and land safely.

38. SpaceX Unveils Mars Vehicle

Source: SpaceX Starship

CEO Elon Musk on to Mars

SpaceX CEO Elon Musk has unveiled his silver bullet: a space-craft to take you to Mars and beyond. It's the new Starship Mk1 prototype, unveiled late in 2019 at SpaceX's South Texas test site in Boca Chico, TX. Starship is a massive, reusable launch system. When finalized, the vehicle will have a 387 foot Starship Super Heavy stack and be powered by six Raptor engines. The spacecraft has room for 37 Raptor engines, depending on the mission needs and distance.

Musk, a Visionary Thinker

Starship will be capable of taking up to 100 people back and forth to space, including the Moon, Mars and other space destinations. Musk says space travel has to be done like air travel. He believes that is the first breakthrough needed to make humans a space-faring, multi-planet civilization. He adds making space travel like air travel is "the fastest path to a self-sustaining city on Mars", which is his ultimate goal and destination.

Passengers in a Year

Musk says the Starship vehicle system could soar into space and back to Earth three times a day. Meanwhile, testing of the vehicle continues with the next test tasking Starship to reach 65,000 feet before returning to Earth. Musk wants the vehicle to successfully perform an Earth orbit by the end of 2020. And, if all goes well, people could start flying as passengers on the vehicle in the next several years.

39. NASA's Shapeshifter Exploring New Distant Worlds

Source: NASA

Drone of Drones

NASA's Jet Propulsion Lab at Caltech has introduced the drone of drones. Right now, it separates into two different units: two shapes to explore new distant worlds and new habitats. It's a prototype and is made of several small, quadcopter drones called cobots. The cobots have propellers and fly independently and the system reassembles to roll on the ground. The future for the concept is much bigger. The Shapeshifter would be made up of a number of small robots that can easily self-assemble into larger robots and disassemble as the mission requires, particularly in space. The mini-robots will be able to fly, roll, float and swim and then morph into a single machine.

Morphing Robots

On the ground, the cobots come together to form a roller-wheel like drone to explore the ground in places in space like Titan, where there is very limited information about the surface. The unpredictability of the surface makes versatility and shape-shifting essential in the drone. In the future, the plan is to make the cobots work together as a team of twelve to explore caves, underwater areas and various types of terrain, including in outer space. The ultimate morphing robot team would be carried aboard a mothership lander that would house their energy source and scientific instrumentation for testing and analysis.

Dragonfly 2026
This extraordinary technology will take quite a few years to fully develop. But a target for it is 2026 when NASA's Dragonfly drone takes off for Titan with possibly the Shapeshifter onboard. The Shapeshifter is a transformational vehicle to explore, treacherous, distant worlds.

40. How to Generate Light at Night

Source: UCLA

Renewable and Complements Solar Power
This new, inexpensive thermoelectric device may be transformative in energy generation. It harvests the coldness of space during the night to generate electricity - enough right now to power a LED light at nighttime, but the inventors say it's very scalable. The device is a significant new innovation from

engineers at UCLA and Stanford University. The gadget works at night when solar systems don't. The inventors say it's a new approach to power generation when power at night is needed. It complements solar power that doesn't work at night, giving a 24/7 approach to green, renewable energy.

Phenomenon Like Frost Formation

The device takes advantage of radiative cooling, the process by which frost forms on grass during above freezing temperatures at night. The sky facing surface of the technology passes heat to the atmosphere as thermal radiation. It loses some heat to space and reaches a temperature cooler than the surrounding air. That temperature differential produces renewable energy at night, when lighting demands are peak.

Scalable Tech for Global Use

According to the UCLA and Stanford engineers, their invention is highly scalable. The radiative cooling device essentially consists of an aluminum disk coated with paint and all the other components are readily available for purchase off the shelf. This is important innovation to watch for because of its practicality and scalability for worldwide use to supplement solar energy at night.

41. Planet Raining Diamonds: NASA Discovery

Saturn Discovery

NASA scientists have discovered that Saturn's mysterious atmosphere is raining diamonds down on the planet. This discovery is extraordinary and yet another example of the vast mineral deposits in space that commercial space ventures are targeting.

Saturn Rings

While studying how and when Saturn's rings of ice and dust emerged, the NASA scientists stumbled upon a surprising feature of Saturn's atmosphere. The atmosphere is composed of high amounts of sulfur, along with hydrogen and helium. The atmosphere is extremely harsh and unforgiving. So much so it rains diamonds onto the planet. NASA's Cassini probe made the shocking discovery when taking a closer look at Saturn's weather in 2013. The discovery was recently released.

From Soot to Diamonds

Saturn is the sixth planet from the Sun and the second largest in the Solar System. According to scientist Dr. Brian Cox, Saturn's atmosphere is brutal. There are huge clouds filled with water and lightning that is 10,000 times stronger than on Earth. The combination changes methane gas in the atmosphere into huge clouds of soot. The pressure is so great the chunks of soot transform into diamonds, which rain down on the planet and liquify because of the intensity of Saturn's environment.

42. China's Exciting Discovery on Moon's Far Side

Source: China Lunar Exploration Project

Unusual, Gel-Like Substance

China's Chang'e 4 lunar rover has discovered an unusually col-ored, gel-like substance in its exploration of the far side of the Moon. The rover named Yutu-2 made the discovery on what the Chinese are calling lunar day 8. Scientists at the Beijing Aerospace Control Center changed the rover's itinerary to focus its instrumentation on the substance found in a small crater. The material's color and luster is unlike anything surrounding it on the lunar surface.

What Is It?

The big question remains what the substance is. Chinese scien-tists haven't offered any explanation. Other experts theorize that it could be melted glass from the collision of a meteorite with the Moon's surface.

Lunar Mystery

Chang'e 4 made the world's first soft landing on the far side of the Moon on January 3, 2019. It's exploring the dark side of the Moon with Yutu-2 and sending back pictures of lunar land-scape never seen from ground level before. Yutu-2, in its travels across the Moon's far side has discovered a lunar mystery with the mysterious lunar gel.

43. Supernova Found in Antarctic Snow

Source: Stock Image Antarctic

Interstellar News

Australian scientists have discovered large amounts of stardust in Antarctica snow, likely from the explosion of a supernova that had been close to the Sun. The explosion rained down particles of a unique iron isotope discovered in melted snow from the South Pole. The discovery of the rare isotope, iron-60, which is not native to Earth, was made by a scientific research team from Australian National University.

South Pole Supernova

The researchers have ruled out any chance that the isotope was created by human activity. They say the only explanation is that it was created by an interstellar rock hitting the Earth. The team melted more than 1000 pounds of snow and analyzed the contents of the melt - large amounts of isotope iron-60. Their next focus is to determine the age of the isotope and to dig deeper for more samples in the icepack. Their remarkable findings were published in the journal Physical Review Letters.

44. Unprecedented Cosmic Event

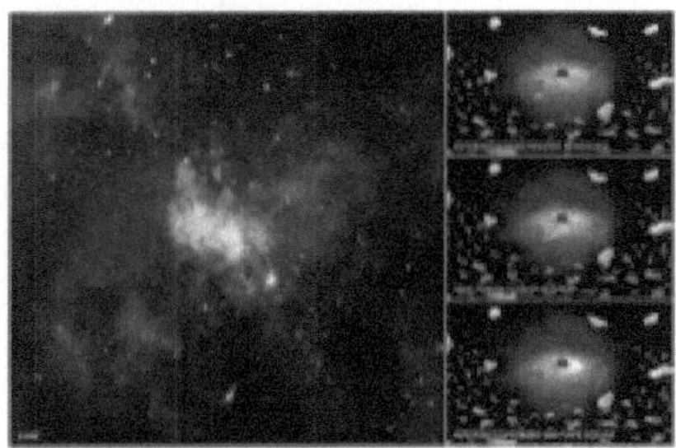

Source: NASA Sagittarius A

Sagittarius A: Black Hole Light Show

The black hole Sagittarius A (Sgr A) is the Earth's closest black hole and is right in the middle of the Milky Way. Using the Keck II Telescope in Hawaii, astronomers watched the massive black hole erupt and spew a huge burst of infrared radiation. They call the event "unprecedented" and they can't explain exactly what caused it.

Off the Charts

The astronomers say the black hole reached much brighter "flux levels" than ever observed before. According to NASA, SgrA is 26,000 light years from Earth. The astronomers observed this space phenomenon mid-2019 and recently disclosed it in the journal Astrophysical Journal Letters. According to the astronomers, black holes are always variable but this one was off the charts of historic data including from the Keck II Telescope.

Theories

There are two theories as to what caused the huge burst of light. A star passing by could have changed the gas flow around the black hole. Or it could have been a flash from a passing gas cloud. The astronomers are anxiously awaiting data from other global telescopes.

45. Passive Cooling from Space

Source: University of Buffalo

Solar Energy and Outer Space Energy

This is new energy innovation from engineers at the University of Buffalo. A green, renewable way to keep buildings cool in urban areas. It's a passive cooling system that absorbs heat and beams it as heat radiation to outer space, while keeping the environment around the system cool. What's remarkable, this system is self-sustaining and requires no electricity or batteries to power it.

Unique and Inexpensive

The device consists of an inexpensive film composed of polymer/aluminum that is put in a box inside of what the engineers call a "solar shelter" that they also invented. The special film absorbs the heat from the air and keeps the surroundings cool. The solar shelter serves two purposes: blocking incoming sunlight and beaming the heat into space through a narrow beam passageway as thermal radiation. The team says that narrow heat radiation passageway is also unique to their system.

Crowded Cities with High Rise Buildings

The narrow beam of heat radiation speeding to space that the system has uniquely created is very important to crowded urban areas surrounded by high rise buildings. It potentially provides a heat thruway to space in crowded places and provides "air conditioned" buildings. To cool a building, numerous units of the system would be needed to cover the roof. The team's landmark research results were published in Nature Sustainability.

46. Martian Rock Tells Big Story

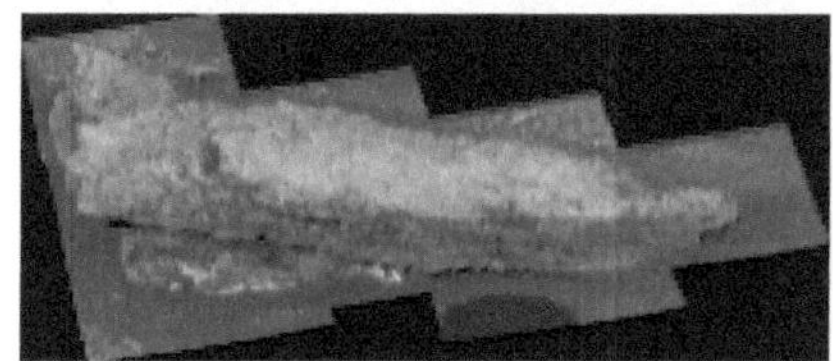

Source: NASA

Shows Presence of Water, Raises Questions of Life on Mars

NASA's Curiosity rover has been exploring the planet Mars for more than six years. It has found a rock that contains some of the secrets to what Mars was like when it was a lot more wet. The discovery is so important that the rock has been nicknamed "Strathdon".

Strathdon

The rock was found in the Gale Crater or the "clay bearing unit". The rock is composed of layers of sediment that are unique and wavy. NASA scientists say this shows evidence of flowing water and blowing wind. They add the history of water on Mars is much more complicated than they had previously thought. They now do not believe that Mars went from wet to dry overnight.

Signs of Life on Mars

The question still remains: with the presence of water on a warmer Mars with a thicker atmosphere, could that have sustained more than microbial life? NASA is searching for that answer.

47. Cubesats in Command

Source: NASA Cubesat

High Theater in Low Earth Orbit

Two cubesats (small satellites) powered by steam have performed a complicated and coordinated maneuver in space. With the help of NASA on the ground, one of the spacecraft commanded the other to close the 5.5 mile gap between them. It did so successfully.

Old Fashioned Steam Power

The twin space crafts' fuel tanks are filled with water. Thrusters convert the water into steam which propels the cubesats. NASA says the maneuvers demonstrate how small spacecraft can work together on future missions. NASA adds these spacecraft are designed to operate in swarms. These space teams could be deployed in the future for deep space exploration and work autonomously to explore new worlds. NASA plans on sending cubesats to the Moon by 2021.

48. Star That's Older Than the Universe

Source: NASA's Hubble Shot of the Star

Is the Universe the New Kid on the Block?

There's a very old star discovered by astronomers that's a cosmic riddle of intergalactic proportions. Astronomers have found a star that they think may be older than the universe. Best estimates by astronomers is that the universe is 13.8 billion years old. But they have found a star close to earth that is estimated to be 14.5 billion years old, based on its very low metal content. The star is HD 140283, nicknamed the Methuselah Star. It's raising a lot of questions, fascination and wonder. Is that beautiful twinkling star older than time and the universe itself?

The Methuselah Star

The questions come from many including a top physicist at the UK's Royal Astronomical Society Dr. Robert Mathews. He calls it a riddle of cosmic proportions. How can the universe contain stars older than itself? This question sparks questions about how accurate calculations are about the age of the universe. Even NASA's estimate is imprecise: Methuselah could be 800 million years younger or older than 14.5 billion years. It is too early to tell but Dr. Mathews is looking into new research into gravitational waves as a possible explanation for the age differential. But, he adds the age of the universe question is back with a vengeance.

49. Charting the Milky Way

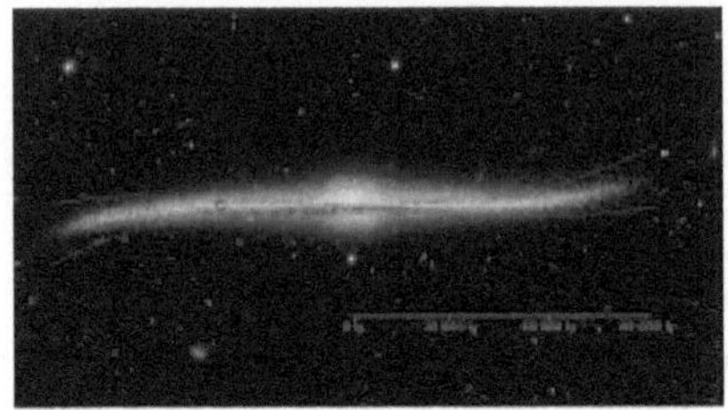

Source: University of Warsaw

Surprise Finding "Warped and Twisted", Not Flat

Polish astronomers from the University of Warsaw have created the most precise and largest map of the Milky Way to date. By tracking thousands of blinking stars in the galaxy, they've created a 3D scale map of the system. They discovered that it isn't flat as a pancake but twists and turns in shape much more than previously thought by astronomers.

Spiral Galaxy

The Milky Way is a spiral galaxy measuring 120,000 light years across. It's a mass of stars, gas and dust about 27,000 light years from the Earth. The Polish team found the distortions in the Milky Way are immense with some stars 60,000 light years away from the Milky Way's center. And, the galaxy's thickness is variable.

Potential Causes

The astronomers cited several potential causes for the variations including interaction with nearby galaxies, intergalactic gas and even dark matter. Their groundbreaking work has been published in the journal Science.

50. SuperEarth Found in Space

Source: NASA's Goddard Space Flight Center

May Have "Earth Like" Conditions

Astronomers have discovered what they're calling a Super Earth 31 light years away from our solar system. They say it is "potentially habitable". The planet is named GJ 357d. It is six times larger than the Earth and it orbits a sun that is much smaller than ours. An international team of scientists discovered it using powerful ground telescopes and following up on data first relayed by NASA's planet hunting satellite TESS. The astronomers believe that it could provide "Earth like conditions" for life.

Looking for Signs of Life

Highly advanced telescopes will, in the next several years, be going on line that will be able to pick out any signs of life on Super-Earth. Two telescopes going live in 2021 and 2025 should show whether the planet is rocky and if it has any oceans. If the planet's atmosphere is thick, it could support water on the surface and sustain life. If there is no atmosphere, the average temperature would be 64 degrees below zero, making it more glacial than habitable.

51. Solar Sailing Through Space

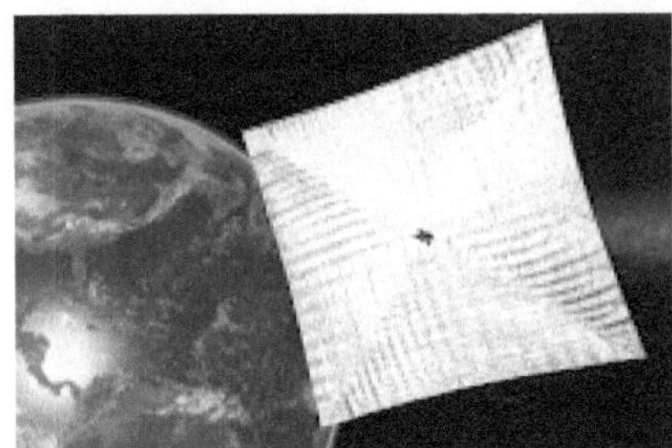

Source: The Planetary Society's LightSail in Orbit

Solar Powered Space Missions

This green energy technology has the potential to be disruptive innovation. It's a solar powered sail invented by the non-profit Planetary Society in California, led by the Science Guy Bill Nye. The LightSail 2 CubeSat spacecraft successfully deployed its solar sail in space. It's the first spacecraft to be propelled by sunlight alone. This is very important science because the use of solar sails to convert solar wind into thrust could save fuel and provide clean solar power for space crafts on long space missions.

52. Solar Sailing

Source: The Planetary Society - image from LightSail

The small spacecraft is the size of a toaster. It was hurled into space aboard a Space X Falcon Heavy rocket in summer 2019. Flight controllers on the ground in California fully deployed the solar sail from the satellite into space. The sail is the size of a boxing ring and the system is now orbiting the Earth solely on the power of the sun. This system is designed to test solar sailing technology and so far it's working very well. It's in full orbit and is relaying pictures back to Earth.

Photon Power

The LightSail 2 is a square composed of four triangles and made of aluminized Mylar. When the sunlight's protons hit the Sail, the pressure creates solar wind and pushes the craft forward. It is targeting to be in orbit through the end of 2020. You can track LightSail 2's progress, location and conditions, on The Planetary Society's website at http://www.planetary.org/blogs/jason-davis/ls2-sail-deployment-live.html

53. NASA's 21 Whole New Worlds

Source: NASA

Planet Hunter Extraordinary

NASA's planet hunter TESS spacecraft - Transiting Exoplanet Survey Satellite - has spent more than a year in space and the results are extraordinary. The spacecraft has discovered 21 new worlds outside of our solar system. It's also spotted 850 possible exoplanets that just have to be confirmed and 6 supernovas. For NASA the results are way beyond all expectations.

Star Search

TESS is focusing on stars that are less than 300 light years away. It looks for dips in brightness which indicates that an object just passed across the star. The data will help NASA determine which exoplanets it wants to explore in-depth and where life might exist.

Historic Mission

This is the most comprehensive search for planets ever accomplished by mankind. TESS finished exploring the southern part of the sky and is examining the north.

54. Sustainable Lunar Bases

Source: Stock Image of Future Lunar Bases

Fascinating Research by the European Space Agency

Administrators of the European Space Agency (ESA) believe that the next step in space exploration is building lunar bases for astronauts. To do that a sustainable source of energy is required to sustain human life and power for rovers and landers. ESA scientists think that they have one, using the surface of the Moon. ESA scientists have created bricks composed of lunar regolith, which is the soil, dust and rocks on the surface of the Moon collected on past lunar missions. In testing, the bricks are able to store solar energy and generate it for use as power, heat and electricity. It's possible this innovation could provide future lunar bases a sustainable energy source.

Scaling Up the Technology

The ESA team is scaling up the process and efficiency of their heat energy bricks from lunar regolith. They say that travelers to the Moon, with this new energy source, wouldn't have to take much with them from Earth. They also think it has the potential to enable very ambitious missions into space. As the world recently marked the 50th anniversary of the Apollo 11 historic mission to the Moon, the work being done by the ESA showcases humanity's continuing quest and ingenuity to journey into space and form habitats there.

55. NASA's Intergalactic Strategy

Source: Artemis on the Moon

Artemis Lunar Program

NASA has announced that it intends to build a long term presence on the Moon. The Agency says that will eventually help humans to get to Mars. NASA has a strategy. The Moon will provide a waystation to Mars and it will serve as a testing ground for the technology needed for interplanetary exploration and habitats.

Sustainable Architecture for Humans in Space

According to the Administrator of NASA Jim Bridenstine, NASA is working to build "a sustainable open architecture" to return humans to the Moon by 2024. He said that NASA is building for the long term on the Moon and then is going to use the Moon as a base to go to Mars and beyond.

Outerspace Technology Testing Site

NASA considers the Moon a testing stage where new technologies can be proven before going to Mars. Once proven, these technologies can help to build self-sustaining habitats on other planets in space such as Mars and beyond. The technologies include harvesting drinkable water for astronauts from subsurface lunar ice. Also, the potential of harvesting from lunar ice, hydrogen and oxygen as rocket propellants developed in space for further interplanetary missions.

56. Robots to Install Telescopes on Moon

Source: Stock image of Moon

Telescopes to Look Deep Into the Universe
A NASA funded lab at the University of Colorado is developing robots to deploy small, highly advanced telescopes on the far side of the Moon. In the next ten years, the NASA team will send a rover aboard a lunar lander spacecraft and place it on the dark, far side of the Moon.

Humans & Machines Working Together
The plan is to set-up a network of small telescopes. The deployment will be performed by the rover's robotic arm. The arm will be controlled by astronauts in an orbiting lunar station called Gateway. Gateway will serve as transit to and from the Moon and as a refueling station for deep space missions.

New Telescopic Views of Deep Space
On the far side of the Moon, the telescope will be free of light

and noise. That will enable a unique and pristine gaze into the deep reaches of space. It's a leading edge example of projects underway by NASA, private companies like Space X and other nations. These missions will open up and change the landscape of the Moon forever.

57. Distant Big Discovery in Space

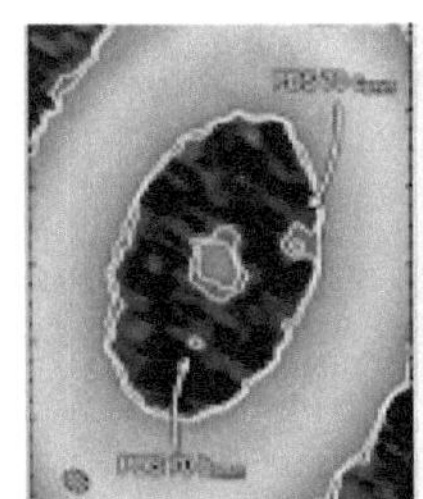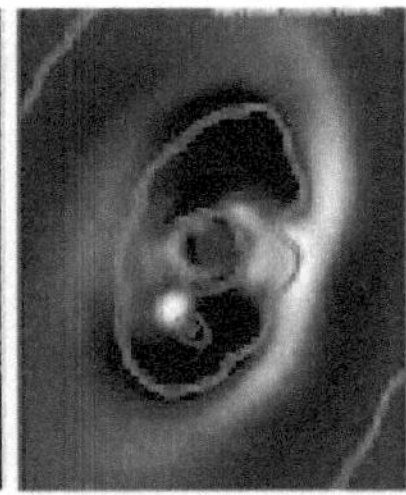

Source: Rice University

World First

Astronomers from Rice University have spotted a moon-forming disk around a very distant exoplanet. The disk, as seen in the millimeter wave radio signal on the left, is made of gas and dirt. Scientists say it is similar to the one that is believed to have created the moons around Jupiter four billion years ago. This is a first time ever observation of the moon forming disk.

World's Most Powerful Telescopes

The Rice University team made the discovery by using the most powerful array of telescopes on Earth. They are located at the ALMA Observatory in Chile.

Exoplanet PDS 70c

The exoplanet is named PDS 70 c. It is a still forming gas giant that's located 370 light years from the Earth.

58. NASA Plans New Space Station

Source: NASA Lunar Orbiter Concept

Man on the Moon in 2024

NASA's plan is for a mini-space station that would orbit the moon with astronauts onboard for 30 to 60 day stays as they travel back and forth to the surface of the Moon. The new space station will be a lunar orbiter, lander and astronaut habitat. It's called the Lunar Orbital Platform - Gateway that NASA wants to use to speed travel and exploration of the Moon. The Trump Administration wants humans on the Moon again by 2024. The Coronavirus pandemic may push that date back into 2025.

Gateway on a Fast Track

NASA wants the 1st phase of Gateway - the Power & Propulsion phase - to launch in 2022. First, NASA will select an industry proposal on phase 1. The 2nd phase is the lunar habitat where the astronauts will live, work and do scientific experiments, on and near the Moon. In 2024, it's expected that both components will be launched on a commercial rocket.

International Partners

Canada has already signed on as a partner on the new space station with NASA. Canada is going to build a robotic arm called Canadarm 3. The European Space Agency is also looking to be a partner. The space station will be equipped to enable the astronauts to do spacewalks. And, NASA envisions that at a later date it will have commercial uses and also be a stopover for astronauts on their way to Mars.

59. From NASA, Clue to Origin of Life

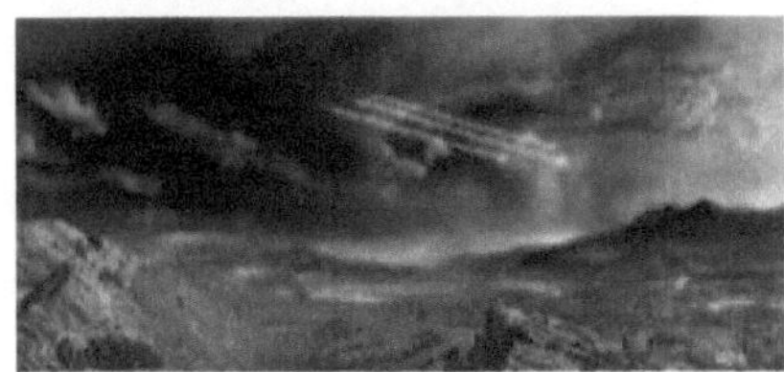

Source: NASA

Cyanide Compounds Found in Meteorites

NASA and a team of university scientists have discovered compounds containing cyanide, carbon monoxide and iron in carbon rich meteorites from space. They think these compounds may have helped power life on earth. They've published their findings in the journal Nature Communications.

Early Earth

The scientists think that cyanide was probably an essential compound for building molecules necessary for life. Besides being present in meteorites, they say the compounds were also present in early Earth before life began when the Earth was constantly bombarded by meteorites.

BENNU

Data collected by NASA's OSIRIS-REx spacecraft of the asteroid BENNU will be delivered to earth in 2023. NASA scientists intend to search for the same compounds which according to NASA scientist Jason Dworkin of the NASA Goddard Space Flight Center "may have helped start life on Earth or on other bodies in the solar system". It's a fascinating innovative research and discovery journey in space to determine the origins of life here on

Earth and perhaps elsewhere.

60. India's Bold Space Goals 2020's

Source: ISRO

Lunar Explorations and Global Cooperation

India is a growing leader in the exploration of space. Of particular interest is India's many lunar exploration successes. India is providing vital data from its multi-year Chandrayaan-2 mission that is now orbiting the Moon. The images are being provided to the NASA-led Artemis program, which is a multi-national effort to put astronauts back on the Moon by 2024. The images of the surface of the Moon are vital to determine the best landing spot on the Moon for the next team of astronauts. Back in 2009, India was the first to discover ice deposits on the surface of the Moon. In 2021, it plans to launch Chandrayaan-3 that will include another landing attempt. Their Vikram lander failed to land safely on the Moon as part of the Chandrayaan-2 mission.

US-India Space Cooperation Accelerating

President Trump in his late February 2020 trip to India praised India's many space exploration and technology accomplishments and pledged growing cooperation on space endeavors with India. NASA and India's space agency IRSO will launch a new satellite in 2022, the NASA-ISRO Synthetic Aperture Radar

Satellite, that will monitor the Earth for floods, glacial changes and soil moisture -- natural events triggered by Climate Change. The US and India are also discussing cooperation on planetary exploration, more Earth observations, heliophysics which is the science of the Sun, human spaceflight and the commercial development of space.

Big Plans for Satellite Launches
India's space plans for the 2020's include large numbers of satellite launches. They will launch 10 Earth observation satellites starting in April 2020. Also they'll launch highly advanced optical, multi and hyperspectral and synthetic aperture radar satellites. Also being launched during 2020 and 2021 are three communications satellites and two navigations satellites.

India In Orbit for the 2020's
India's space plans are big and ambitious. They are vying for a leadership role alongside the US, Europe, China and Russia. Just for 2020, they have a budget of $2 billion, big accomplishments, big plans and big ambitions.

India's Gaganyaan Human Space Flight Program
A big piece of technology for India's space ambitions is its GSLV Mk111 rocket. An unmanned test flight will take place in December 2020 and a second in July 2021. India wants to send a three member crew into a low Earth orbit for 5 to 7 days. They are also working on a reusable launch vehicle with the aim of developing a winged body vehicle similar to an aircraft.

61. Out of This World Exoplanet

Source: NASA Artist Rendering

Exoplanet LHS 3844b

A NASA space telescope has revealed for the first time a rocky Earth-sized exoplanet. It's outside of our solar system and tightly orbits around the most common type of star in our home galaxy, the Milky Way. The exoplanet has no atmosphere to sustain life.

Atmospheric Void

Researchers published their discovery in the journal Nature. They say the surface of the exoplanet, known as LHS 3844b, is most likely barren like the Moon or covered in dark volcanic rock. The exoplanet is 49 light years away from the Earth. It's one of 4000 exoplanets, discovered in the past 20 years, that orbit stars in the Milky Way. It's 1.3 times larger than the planet Earth.

62. UAE's Hope Mars MIssion

Source: Emirates Mars Probe/University of Colorado

United Arab Emirates Going to Mars

The United Arab Emirates has big space ambitions and plans. It is going to launch a mission to Mars in 2020, making it the first Arab nation to launch an interplanetary mission. The Hope Mars probe will deploy the Hope Satellite to study the climate on Mars.

Destination Mars 2021

When the Hope satellite reaches Mars in 2021, it will show how Mars' climate conditions change throughout the year. The satellite will be launched from Japan in June 2020. It will reach Mars in February 2021 and collect data for two years. The mission could be extended through 2025. Interesting, a female scientist is a key player behind the mission. Sarah Al Amiri is deputy project manager of the Mars Mission and Chairperson of the UAE Council of Scientists. 90% of the workers on the Emirates Mars Mission are under 35 years old.

Highly Advanced Technology

The Mars probe will carry three highly advanced pieces of equipment. There's the Emirates Exploration Imager, which is a camera that will send high-resolution images back to Earth. The Emirates Mars Infrared Spectrometer will study ice, water vapor, dust and temperature patterns in the Martian atmosphere. And, the Emirates Mars Ultraviolet Spectrometer will study Mars' lower and upper atmosphere and be used to determine what causes hydrogen and oxygen on Mars to escape into space.

Crowded Martian Skies

The UAE Mars team is working with space experts at the University of California Berkeley, the University of Colorado Boulder and Arizona State University. Mars in the 2020s will be a crowded place with four major space missions underway. NASA's 2020 Rover mission, the European Space Agency's Exo-Mars rover mission, China's Mars Explorer and now the Emirates Mars Mission will all be probing Mars, which is 140 million miles from the Earth.

63. Trump Accused of Planet Snatching Plan

Source: NASA

Russia & US Intergalactic War of Words
This could be a script for a new George Lucas "Star Wars" movie. But it's not science fiction. It's global news on space. The Russian Space Agency Roscosmos has accused President Donald Trump of creating a basis for
snatching planets. At issue is an Executive Order the President signed detailing US policy on commercial mining in space. From Russia's perspective, the Executive Order slams the concept that space is the property of all of humanity.

At Stake - Intergalactic Money in Trillions of Dollars
Experts believe that the mineral deposits that space

contains could be worth trillions of dollars. Mr.
Trump's Order states that the US will negotiate
bilateral and multilateral statements and agreements
on "safe and sustainable operations for the
public and private recovery" of resources in space.
It adds that the US does not "view space as a
global commons." It appears that Russia's fundamental
problem is the US is attempting to lead the effort.

Russia Takes Intergalactic Offense

Russia's space agency is sounding the alarm
that the US might "attempt to expropriate outer
space and implement aggressive plans to actually seize terri-
tories on other planets". Russia says any attempts to privatize
space are unacceptable.

US Position on Space Resources

The White House says it is trying to establish
international support with the US position that
space resources can be used by companies and
organizations. A US federal law grants US companies rights to
the space resources they extract. To be more specific, the US
is encouraging international support for the public and private
recovery and use of resources they discover and extract in outer
space.

NASA's Big Lunar Mission

A key reason behind the President's Executive Order
is to support NASA's long term plans for exploration of the
Moon. In April 2020, NASA outlined its plans for sustainable
exploration of the Moon. The plans include the development
of what it calls "in-situ" resource utilization technologies to use
water ice and other lunar materials found on the Moon to create
fuel, oxygen and other materials for sustainable Moon surface
operations. These lunar operations would require decreasing
amounts of supplies needed from Earth.

Commercializing Outer Space

This is a brand new chapter in Space that's becoming very relevant. It's about commercializing the exploration and discovery of mineral assets in space, that superpowers, including China, are laying claims to. Space is a huge, new commercial opportunity and global superpowers are vying for their piece of it.

"SPACE: THE UNKNOWN REGIONS" © BY EDWARD KANE ON AMAZON

Copyright 2020 © Edward Kane

ALL RIGHTS RESERVED

BOOKS BY JOURNALIST EDWARD KANE

Non-fiction books on innovations and discoveries across industries in 2020 and 2019.

TOP INVENTIONS FOR THE 2020's
Just Published by Amazon and Kindle amazon.com/author/ekane

SMART DEVICES FOR THE 2020's

Just published by Amazon and Kindle amazon.com/author/ekane

FUTURE TRAVEL VEHICLES

Just published by Amazon and Kindle amazon.com/author/ekane

FUTURE TRAVEL VEHICLES BY EDWARD KANE

SPACE

- "Space 2020's: What's Up There?" - Kindle & paperback
- "Bargain Space Trips" - Kindle & paperback
- "Space Renaissance in the 21st Century" - Kindle & paperback
- "Search For Life in Space" - Kindle & paperback

TRANSPORTATION AND TRAVEL INNOVATIONS

- "Future of Transportation: 2020's and Beyond" - Kindle & paperback
- "Electric Vehicles for All" - Kindle & paperback
- "Hot Electric Vehicles for the 2020's" - Kindle & paperback
- "Important Innovations: Transportation" - Kindle, paperback & Audiobook on Audible
- "How to Travel in the Future" Vol. 1 & 2 - Kindle & paperback

LISTS OF TOP NEW INNOVATIONS

- "List of Best New Innovations" - Kindle & paperback
- "List of Top New Environmental Innovations" - Kindle & paperback
- "List of Top New Gadgets" - Kindle & paperback
- "List of Top New Medical Innovations" - Kindle &

paperback
- "List of Top New Energy Innovations" - Kindle, paperback & Audiobook on Audible
- "List of Top New Robots" - Kindle, paperback & Audiobook on Audible
- "How to Use AI & AR" - Kindle & paperback

INVESTING IN INNOVATIONS

- "Investing in Disruptive Innovations" - Kindle & paperback

Fiction - Adventure, Life Lessons Book for Children

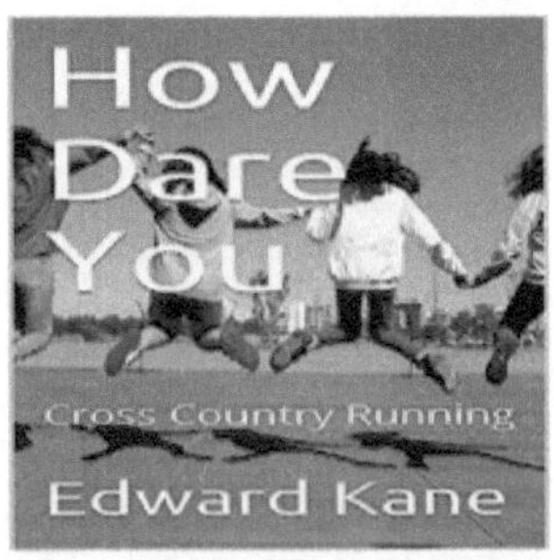

- "How Dare You" - Kindle, paperback & Audiobook on Audible

amazon.com/author/ekane